JIT Implementation Manual

The Complete Guide to
Just-in-Time Manufacturing

Second Edition

Volume 3

JIT Implementation Manual

The Complete Guide to
Just-in-Time Manufacturing

Second Edition

Volume 3

Flow Manufacturing –
Multi-Process Operations and Kanban

HIROYUKI HIRANO

CRC Press
Taylor & Francis Group
Boca Raton   London   New York

CRC Press is an imprint of the
Taylor & Francis Group, an **informa** business

A PRODUCTIVITY PRESS BOOK

Contents

Volume 3

Volume 4

Volume 5

Volume 6

Chapter 5

Flow Production

Why Inventory Is Bad

Why Does Inventory Accumulate?

Every year, when heavy rains hit the forest, the streams and rivers suddenly swell and sometimes overflow. Most river flooding is caused by localized downpours. The rivers become wider and sometimes adjacent forks are reunited as a single large river.

In factories, goods and materials should flow in the factory much as water flows in a river. But things tend to accumulate. We could say that the "river"—the flow of in-process inventory—tends to "flood." Needless to say, it would be better if this river of in-process inventory flowed smoothly and briskly. The following are some of the main reasons for such "flooding" in factories.

Reason 1: Inventory flow is behind the times

It has been a long time since large lot production gave way to the era of wide-variety, small lot production, but some manufacturers still have not caught up. They try to use the old "shish-kabob" production schedules to turn out orders for a wide assortment of product models in small lots and, not surprisingly, "floods" often occur at certain processes.

Reason 2: Old habits are hard to change

Some factory managers understand quite well that this is the era of wide variety and small lots. But they do not have the energy and courage to let go of old familiar ways and make the necessary changes. Rather than trying to "go with the flow," they are just trying to stay afloat for the years remaining until their retirement age.

Reason 3: Unbalanced capacity brings unbalanced inventory

Inventory shoots through the "rapids" of high-capacity processes, but it naturally gets backed up when it reaches processes having lower capacity.

Reason 4: Inventory is sometimes gathered from several processes

Some processes, such as painting and rinsing processes, often use large equipment that can handle in-process inventory sent from several processes. Naturally, the in-process inventory from several processes accumulates at such large equipment before being processed by it.

Reason 5: Inventory must wait to be distributed from large processes

This is what happens at the downstream side of the large equipment described under Reason 4. Each kind of processed inventory must wait its turn to be sent on to one of several downstream processes.

Reason 6: Inventory must wait for a busy operator

Sometimes operators work sequentially on a number of machines. We call this "caravan" operations. In-process inventory tends to gather at each machine until the operator gets a chance to process it. In other words, inventory gathers wherever the operator is not.

Reason 7: Inventory accumulates when operators dislike changeovers

Inventory tends to gather at presses and other processes where changeover is regarded as arduous work. The operators would much rather do fewer changeovers by handling large lots.

Reason 8: Inventory accumulates in factories that have "end-of-the-month rushes"

This tends to happen at factories that have monthly volumes to meet. The assembly line is especially busy during the last five days of the month. In fact, workers from all over the factory are called over to the assembly line for the end-of-the-month rush. By the middle of the month, the factory is chock-full of in-process inventory, lined up to be assembled during this rush period.

Reason 9: Inventory accumulates due to faulty production scheduling

Sometimes production planners are not knowledgeable enough about inventory and include some noninventory items as inventory. Such misunderstandings can lead to incorrect inventory distribution planning when drawing up production schedules.

Reason 10: Inventory accumulates when people forget to revise standards

Once standards are set for lead-time, lot sizes, or acceptable defect rates, people forget to revise them. Soon workshops start producing extra goods in anticipation of a certain percentage of defectives. Surplus production means surplus inventory.

Reason 11: People tend to store up "just-in-case" inventory

Things do not always go as planned. Sometimes, new developments in a company's business activities will require a sudden change in production scheduling. All company divisions—from sales to management, purchasing, and manufacturing—like to keep a "safety margin" of extra inventory around just in case a sudden change of plans occurs. "Safety" is a misleading term here. What these inventory buffers provide is not safety, but security for the people in charge.

Reason 12: Inventory accumulates due to seasonal adjustments

No product sells at the same rate all year-round. Some sell in cycles, and others have distinct seasons. No one in factories likes to deal with sudden and dramatic changes in production. Instead, they try to smooth out the seasonal transitions by producing ahead of time in anticipation of extra orders when the product's season arrives. Obviously, this requires some stockpiling of inventory.

Thus, there are at least a dozen major reasons why inventory tends to accumulate in factories and throughout entire companies. Unless the company's various departments come to grips with these reasons, inventory will keep on building until it begins to sap the company's strength.

Why Is Inventory Bad?

Most people regard inventory as a "necessary evil." They feel especially strong about an inventory's necessity when sales are brisk, but when sales sag inventory starts looking evil. So it is a necessary evil—necessary today and evil tomorrow.

While most Western companies tend to look upon inventory as a necessary evil, most Japanese companies emphasize its wickedness. In fact, attitudes toward inventory is one key characteristic of the difference between Western and Japanese manufacturing systems.

In Japan, inventory is regarded as being so evil that it is often called "the company's graveyard." Japanese managers tend to view inventory as the root of all evil and a likely cause of poor performance in any business activity.

But why is inventory so evil? Again, there are several reasons:

Reason 1: Inventory adds to the company's interest payment burden

Inventory solidifies a lot of capital (as inventory assets) that could otherwise be turned over for a profit. It puts pressure on operating capital and raises the interest payment burden. Therefore, it is clearly an obstacle to successful business management.

Reason 2: Inventory incurs maintenance costs

Inventory is an investment of capital that does not of itself contribute to profits. Moreover, inventory has to be managed and maintained, which adds to costs: warehouse lease fees, insurance premiums, property tax, and so on.

Reason 3: Inventory means losses due to hoarded surpluses and price cutting

When there is excess inventory, unused items undergo age-related deterioration. They get hoarded up due to their obsolescence or they are sold off at rock-bottom prices, both of which hurt corporate profitability.

Reason 4: Inventory takes up space

Naturally, any inventory we have takes up a certain amount of space. Eventually, the piles of inventory start spilling over into the warehouse aisles, which leads to building new shelves and even a new warehouse.

Reason 5: Inventory causes wasteful operations

Inventory causes goods to be retained. Retained goods always require some kind of conveyance. Conveyance never adds value to the product. Warehouse operations include picking up, setting down, counting, and moving—none of which add value (therefore, all of which are wasteful).

Reason 6: Inventory requires extra management

Warehouse operations need to be managed. Managers have to keep track of when items are received at the warehouse, when they are shipped out, and the current amount of each item in the warehouse.

Reason 7: Inventory requires advance procurement of materials and parts

Companies that keep large warehouses make it a practice to order materials and parts even before client orders come in. These parts and materials, however, do not always match what is actually required by the orders.

Reason 8: Inventory incurs wasteful energy consumption

Building, operating, and managing warehouses means greater energy costs incurred by electric, pneumatic, and hydraulic equipment.

These eight are just the more obvious reasons why inventory is bad. We have not even begun to consider other reasons related to capital turnover, hoarding surpluses, and the like.

What, more than anything else, makes inventory evil? This question deserves some sober contemplation. Let us look at a few of the reasons that we have not yet covered.

First, there is the greater interest payment burden incurred by inventory. Let us assume that a certain company has plenty of money, and does not need to worry about paying interest. The managers at this company see no harm in having several warehouses for its factory. "Hoarding up surpluses" is a problem at these warehouses, but the managers think the way to solve this problem is by making products that tend to sell briskly.

Let us reconsider the problems caused just by taking up space. In a huge warehouse, wasted space is rarely noticeable. If anything, we would get the feeling that not making use of the immense warehouse is somehow wasteful. But the real waste lies in having such a large facility to begin with. No matter how much capital a company has, no matter how quickly its products sell, and no matter how much space its factory sites include, inventory remains just as evil a thing as ever.

So what might we say is the *real* reason why inventory is bad? I have found this most basic reason is: Inventory conceals all sorts of problems in the company.

There are a countless number of factories in the world. Each factory must deal with a wide variety of problems every day. Problems pile up even at the best factories, and *there is no such thing as a problem-free factory*.

Let us compare problems in factories to rocks that pile up at the bottom of a pond. When the pond is full of water, we do not see any of the rock piles, but if we empty the pond, they suddenly become obvious. Figure 5.1 illustrates this analogy.

Keeping a large inventory of finished products in the warehouse enables the company to deal with the demands of

• High water volume (inventory volume) conceals the rocks (problems)

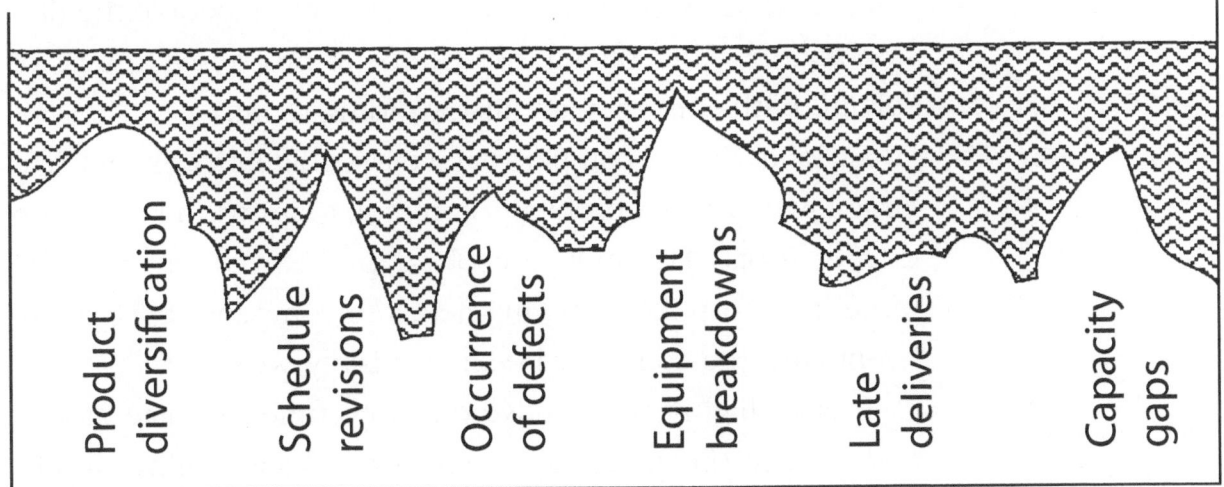

• Low water volume (inventory volume) reveals the rocks (problems)

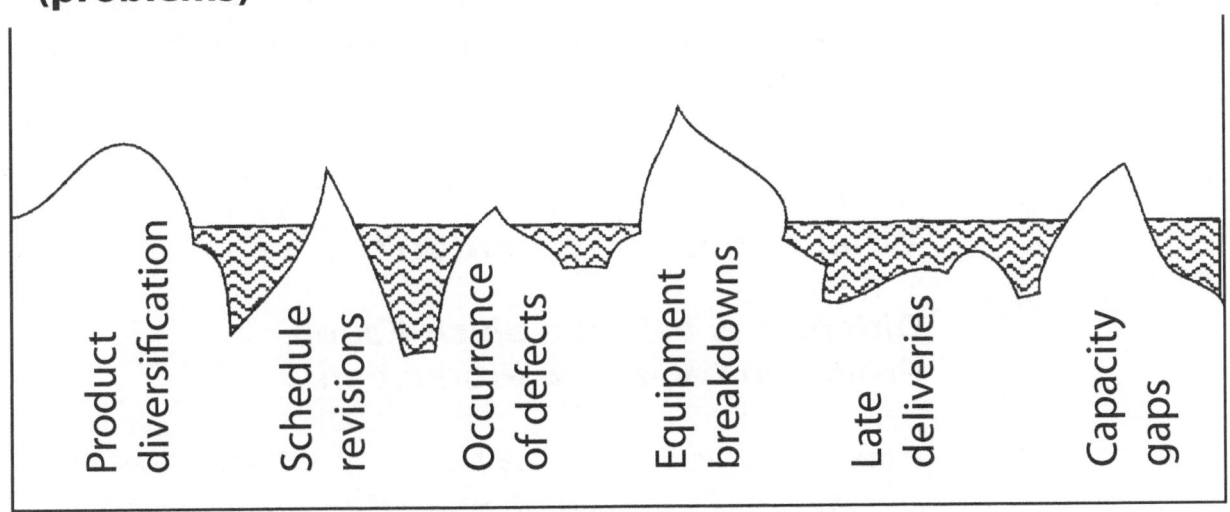

Figure 5.1   How Inventory Conceals Various Problems Affecting the Company.

product diversification without having to address the problem of why it takes so long to switch production from one product model to another. It also enables the company to keep up with schedule changes without having to question why schedule changes are so hard to keep up with in the first place. Plentiful warehouse supplies can also help fill in the production output gaps caused by equipment breakdowns, again without having to take preventive action against the problem.

In short, a "well-stocked warehouse" gives people the illusion that they are solving these kinds of problems. Instead of solving problems, they are just avoiding them.

As long as the company avoids problems by keeping a large inventory, the problems continue to grow and lay down deeper roots. The more unsolved problems there are, the more inventory the company needs to compensate for them. Eventually, the company becomes visibly weaker.

Today's highly competitive era is no time to waste money and energy on covering up problems. Challenging trends, such as product diversification and shorter delivery deadlines, create new problems every day. The successful companies are the ones who not only learn how to respond rapidly to today's fast-changing marketplace, but also know how to apply the same swiftness in dealing with problems—not avoiding them.

What Is Flow Production?

Differences between Shish-Kabob Production and Flow Production

I mentioned earlier that the factory "river"—the flow of in-process inventory—tends to "flood." A main reason for such flooding is conventional lot production, which we might also refer to as "shish-kabob production." The shish-kabob image is a natural one—workpieces move along in little clumps. In other words, they are grouped into batches for batch processing at each workshop along the line. We can look at the differences between shish-kabob production and flow production in various ways (see Figure 5.2). Let us look at some of these in more detail.

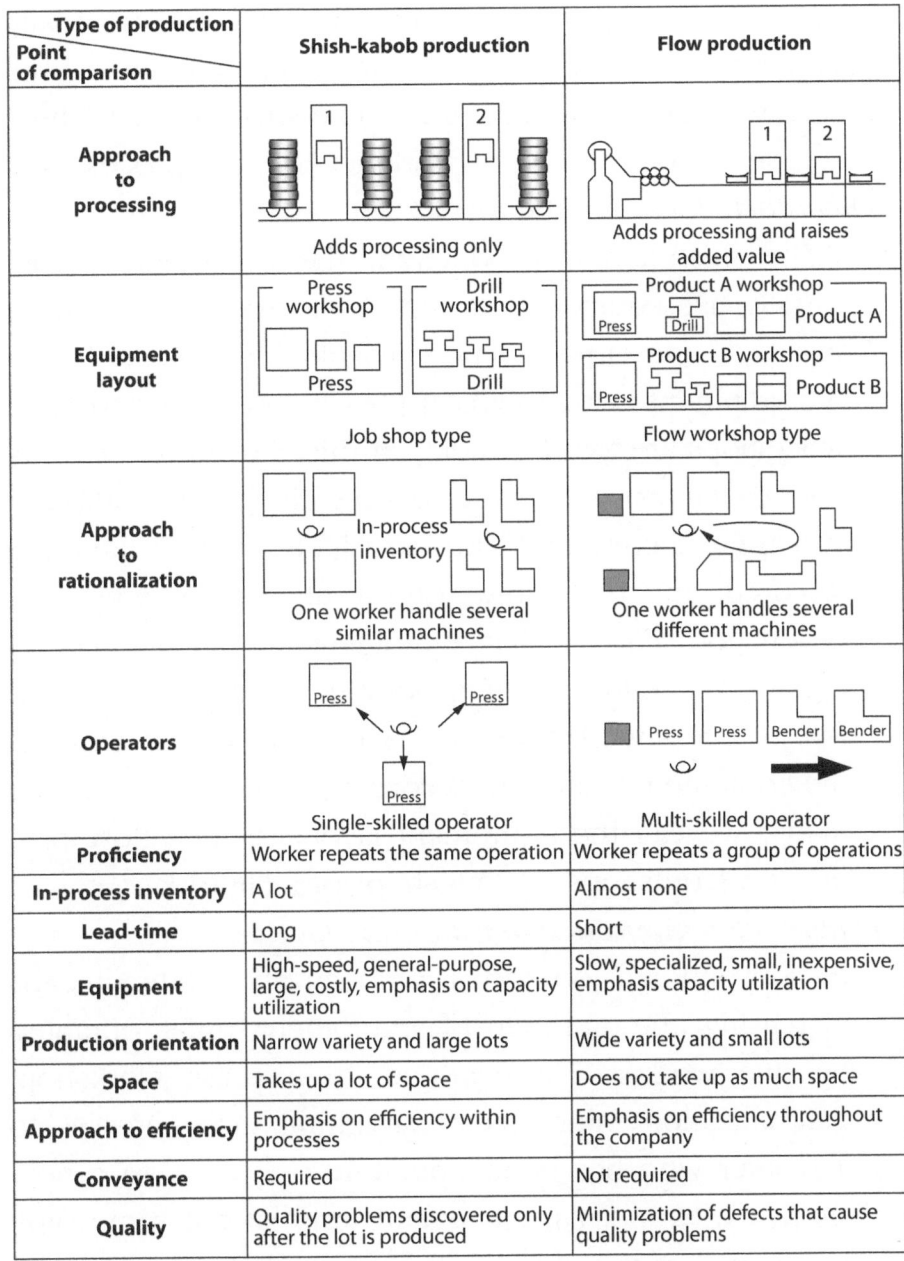

Type of production / Point of comparison	Shish-kabob production	Flow production
Approach to processing	*Adds processing only*	*Adds processing and raises added value*
Equipment layout	*Job shop type*	*Flow workshop type*
Approach to rationalization	*One worker handle several similar machines*	*One worker handles several different machines*
Operators	*Single-skilled operator*	*Multi-skilled operator*
Proficiency	Worker repeats the same operation	Worker repeats a group of operations
In-process inventory	A lot	Almost none
Lead-time	Long	Short
Equipment	High-speed, general-purpose, large, costly, emphasis on capacity utilization	Slow, specialized, small, inexpensive, emphasis capacity utilization
Production orientation	Narrow variety and large lots	Wide variety and small lots
Space	Takes up a lot of space	Does not take up as much space
Approach to efficiency	Emphasis on efficiency within processes	Emphasis on efficiency throughout the company
Conveyance	Required	Not required
Quality	Quality problems discovered only after the lot is produced	Minimization of defects that cause quality problems

Figure 5.2 Comparisons of Shish-Kabob Production and Flow Production.

Difference 1: Approach to processing

Shish-kabob production uses large groups of work-pieces at each processing point within a process station. These groups (lots) are retained at the process until all of the units in the lot are completed. By contrast, flow production means that once each workpiece has been processed, it is sent to the next process for immediate processing. This continuous moving flow continues until each workpiece is completed as a finished product. There is little or no retention of workpieces at the processes.

Difference 2: Equipment layout

For shish-kabob production, the equipment layout usually has equipment grouped into rows of machines that serve the same function. This is the "job shop" type of equipment layout. Typical press workshops and lathe workshops are two examples of this. Since flow production means processing and sending along one workpiece at a time, there should be very little material handling required, and preferably none at all. That is why flow production requires that equipment be laid out according to the sequence of processes. Workshops are no longer "press workshops" or "lathe workshops." Instead, the equipment is laid out according to the product being made. We call the equipment layout in such flow production workshops a "flow shop" or a "line" layout.

Difference 3: Approach to rationalization

In conventional job shops, rationalization often means increasing the number of equipment units operated by one worker. For example, in a press workshop, rationalization might mean assigning three presses to a worker who has been operating only two. In a flow shop, we cannot assign several units of the same type of equipment to a single worker, since that would interrupt the one-piece flow of workpieces from process to process. Instead, individual workers learn to operate several different kinds of equipment corresponding to the different processes along the line. We call this "multi-process operations." (For a more detailed description of multi-process operations, see Chapter 6.)

Difference 4: Operators

No matter how many equipment units each worker operates in conventional job shops, the worker sticks to a single set of skills as a press operator, a lathe operator, or whatever. In flow shops, workers learn several sets of skills needed to operate a series of different processes, such as press → drilling → bending. We call such workers "multi-process workers."

Difference 5: In-process inventory

In the shish-kabob production system, in-process inventory is found as lots retained between processes and between machines. In flow production, where workpieces continually flow from one process to another, there is rarely any in-process inventory retained between processes or machines.

Difference 6: Lead-time

Shish-kabob production tends to create long lead-times because of the many times when lots are retained while waiting for the previous lot to be processed or for the rest of the same lot to be processed. When flow production keeps workpieces moving all the way until the final process, the lead-time can be reduced to the level of the total processing time.

Difference 7: Equipment

Shish-kabob production lacks any kind of overall flow from raw materials processing to final product assembly. This makes it very difficult to sense rhythm in the factory operations. The only kind of rhythm that might be evident is the pitch at which individual workers operate individual machines. This is called the "individual rhythm." Shish-kabob production managers seek to improve factory operations via *greater speed,* which requires *general purpose machines* that can quickly process various types of workpieces. However, general purpose machines tend to be *large* and *expensive.* When large and costly machines are installed, the factory managers naturally become concerned with maintaining *a high capacity utilization rate* by turning out more and more products. Meanwhile, the factory becomes one that is more concerned with its equipment than with its customers.

Flow production takes an almost completely opposite approach by emphasizing a smooth production flow all the way from materials processing to final product assembly. There is a clear overall rhythm to production, and the tempo of this rhythm is set by customer orders. Each machine along the production line is like a bar of music. There is no need to hurry the tempo. Production should always be *slow enough to remain in the overall flow*. There is also no need to hurry when changing over to other product models. Each machine should serve only one main function, operating like a bar of music in the symphony of production. Each machine should be a *specialized machine* that emphasizes quality over speed. These specialized machines should serve only the minimum required function and should be *compact* enough to fit right into the production line. Naturally, these slower, more specialized machines are *inexpensive* and therefore do not invite concern over capacity utilization rates. Instead, the major maintenance concern is to ensure a *high possible utilization rate* (that is, high serviceability) to prevent disruptions in the production flow.

Flow Production within and between Factories

"Flow" can mean the gurgling flow of tiny brooks amid the rocks or the quiet majestic flow of a wide river spanned by long bridges. In the factory, the smaller parts lines are like the brooks and the large final assembly lines are like the wide rivers. The streams eventually converge into rivers, and the flow (of goods) ultimately reaches the sea (the marketplace).

Factories need to have a smooth flow of operations, and the basic method for creating such a flow is by making individual improvements. These improvement "points" add up until they form a "line" of improvements. This line is the flow between processes.

Eventually, we also need to have a smooth flow of production operations between manufacturers and the vendors, subcontractors, and wholesalers or distributors that they work with. This kind of flow is a vertical flow between factories, and the corresponding improvements are called vertical improvements.

Therefore, when we discuss flow production, we must be aware of the kind of flow production we are talking about. The main distinction to make is between flow production within a factory and flow production between a factory and another factory or business.

1. *Flow production within a factory.* To establish this kind of flow production, we must eliminate the in-process inventory that accumulates at and between processes as "flood water" or "shish-kabob clumps."
2. *Flow production between factories.* We must also establish a smooth flow of operations between our own factory and the various subcontractor factories, vendors, distributors, and other businesses that our factory deals with.

Flow Production within the Factory

Eight Conditions for Flow Production

Making things requires various techniques. Many of the techniques used in manufacturing are based on two engineering technologies: pressing and drilling (or punching).

So we might ask whether JIT improvement is meant to also improve these essential engineering technologies. The answer is yes. JIT improvement means radical improvement, which means it goes into the very basic engineering technologies. But that is not the main point of JIT improvement.

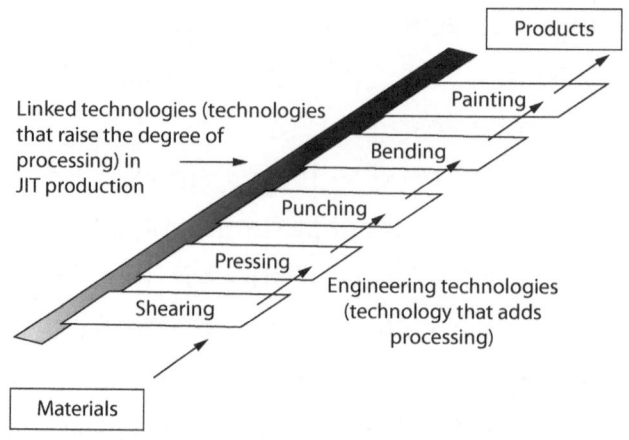

Figure 5.3   Linked Technologies in JIT Production.

The engineering technologies, such as pressing and drilling (or punching), are technologies for processing workpieces.

Of course, no matter how many times a press adds processing to a workpiece, it will not be enough to turn out a finished product. Manufacturing products requires an assortment of materials plus several engineering technologies, among which pressing is just one.

The main work of JIT improvement is to link these engineering technologies in a production system that is attuned to customer needs. (See Figure 5.3.)

While engineering technologies add processing to workpieces, linked technologies raise the degree of processing. Accordingly, the basic aim of JIT production is to make things one at a time, in a smooth flow, to prevent defects.

The following is a list of eight conditions that must be met to establish one-piece flow production.

Condition 1: One-piece flow

Condition 2: Lay out equipment according to the sequence of processes

Condition 3: Synchronization

Condition 4: Multi-process operations

Condition 5: Training of multi-process workers

Condition 6: Standing while working

Condition 7: Make equipment compact

Condition 8: Create U-shaped manufacturing cells

Condition 1: One-Piece Flow

One-piece flow is the most basic of all eight conditions; it is where flow production starts and ends. One-piece flow refers to the condition in which each workpiece must be processed and passed along the production line by itself, and that includes assembled quasi products. One-piece flow sounds simple enough in theory, but putting it into practice can be very difficult indeed.

Whenever we inspect the production line and find places where "shish-kabob clumps" of in-process inventory have accumulated, we need to find out why it happened. Perhaps the equipment units are not lined up according to the processing sequence, or perhaps the processes are not synchronized. There is always some reason, and it usually includes a human factor: resistance to change. That is why it is so important that *everyone* understands what JIT is about from the outset. Without prior understanding, things are bound to fail.

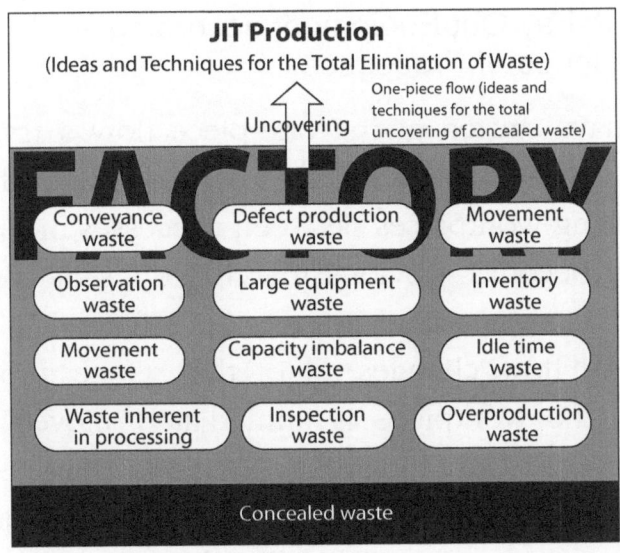

Figure 5.4   One-Piece Flow.

JIT production means ideas and techniques for the total elimination of waste. We must begin by uncovering all of the deeply rooted concealed waste in the factory. Switching to one-piece flow is the best way to do this. If I may paraphrase the JIT definition: *One-piece flow means ideas and techniques for the total uncovering of concealed waste.* (See Figure 5.4.)

Unfortunately, one-piece flow is not something we can achieve simply by rearranging the equipment according to the processing sequence and retraining the workers in new operation procedures. Rather, it is a first step in a process that includes uncovering concealed waste in the factory. That is why we should begin by switching over to one-piece flow *using the current equipment layout and operation procedures.* This will show us where the hidden waste is, such as conveyance waste, waste caused by having large equipment, and so on. Once we have uncovered all of this waste, we are more than halfway there since we have learned how to redesign the layout to eliminate the conveyance waste (by eliminating conveyors), large equipment waste (by using only compact equipment), and other waste.

The key to success in all of this is whether or not we are truly resolved to implement one-piece flow production.

Condition 2: Lay Out Equipment according to the Sequence of Processes

After we have started giving one-piece flow a try, we first notice conveyance waste staring us in the face. If the line was conveying workpieces between processes in lots of 100, it suddenly becomes obvious that 100 units of conveyance waste had been concealed in each lot.

One-piece flow changes all of that. Once a process is completed, the workpiece is immediately moved along to the next process. Under current conditions, that means each workpiece must be moved along via the existing conveyance system. The amount of time and trouble built into that system suddenly becomes 100 times greater. That makes it obvious enough for the workers to notice the tremendous amount of waste involved. With that awareness, they are ready to start changing the equipment layout.

In redesigning the equipment layout, they now know the idea is to minimize conveyance or, better yet, eliminate it altogether. They can do this by lining up the equipment according to the processing sequence. This kind of line up is the standard for all flow shops and flow-oriented production lines.

Condition 3: Synchronization

Once we have set-up the equipment for flow production, we need to consider how fast the flow should be; in other words, at what pitch the processes should be operated. Unless we have a common pitch among processes, workpieces will accumulate at the slower processes and cause the flow to "flood."

Synchronization means maintaining the same pitch among the various processes. In the final analysis, the pitch should be determined (as so many minutes and seconds) by the amount of orders from customers. This time figure is called the cycle time. The cycle time sets the rhythm for the "music" of manufacturing. (Cycle time is discussed in more detail in Chapter 10 of this manual.)

Condition 4: Multi-Process Operations

One-piece flow production can be achieved without any multi-process operations. (See Chapter 6 for further description of multi-process operations.) Instead, we can simply assign one worker to each process and have them process and hand along workpieces according to the established pitch. Figure 5.5 illustrates this kind of arrangement, which we might call "hand-transferred one-piece flow."

One problem with the hand-transferred one-piece flow arrangement is that requiring one worker at each process makes it difficult to add or subtract workers to adjust for changes in scheduled output. Such adjustments are the aspect of JIT known as "manpower reduction" (described in Chapter 7).

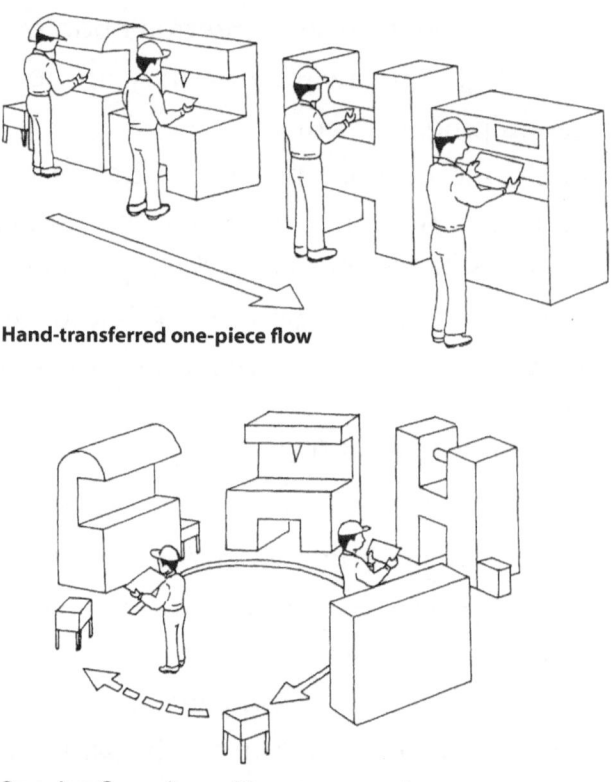

Hand-transferred one-piece flow

One-piece flow using multi-process operations

Figure 5.5   Two Types of Flow Production.

The idea is to have just the minimum amount of manpower needed to produce the scheduled amount of output.

Another problem with the hand-transferred one-piece flow arrangement is that it encourages workers to think of themselves in strictly defined job roles, such as press operator, drill operator, or inspector. This reduces manpower flexibility and makes it hard for the idea of "building quality in at each process" to take hold among the workers.

These are two reasons why JIT production calls for flow production using multi-process operations. Multi-process operations move vertically along the production line by having workers operate as many processes as possible. This is quite different from multi-unit operations, in which workers expand their work horizontally in the production line by operating several of the same type of machines performing the same process.

Condition 5: Training of Multi-Process Workers

Multi-process workers are workers trained to handle several processes together. Conversely, we call workers that handle only one process "single-process workers." (See Chapter 6 for a detailed description of multi-process workers.)

Training multi-process workers is a key step toward achieving JIT flow production. This training can be extended company-wide over the short term to include:

■ Thorough standardization of machines and other equipment so that anyone can more easily learn to operate them;

■ Equally thorough standardization of operations, eliminating special or exceptional cases;

■ Company-wide multi-skill training as an important part of company-wide improvement.

Condition 6: Standing While Working

In most machining workshops, workers traditionally stand while working. However, assembly lines such as at home electronics and electrical equipment manufacturers are usually operated by workers who sit while working. The switchover to standing while working can create serious problems at such places. It may take a long time indeed before such assembly workers are convinced of the need to stand while working. (One wonders if it might even take as long as it took our primeval ancestors to switch from walking on all fours to walking on their legs only!)

About the only way to succeed in this difficult transition and overcome workers' reluctance to stand is by getting the entire company deeply involved—including the president and other top managers—in pointing out the advantages that standing while working brings, i.e., easier movement, helping each other out when necessary, correction of unbalanced operations, multi-process operations, and much more.

Condition 7: Make Equipment Compact

If one workpiece is about as big as a fist, then a lot of ten workpieces would be about the size of a bread box and a 100-workpiece lot would be as large as a washing machine.

To handle lots of 100 workpieces each, we need a conveyor that can easily move washing machines. Likewise, the processing machines and other equipment must also be able to handle washing machine-size lots.

In other words, the equipment has to be big, so big that much of it will not fit into a small production line. In most cases, we must set such large equipment aside somewhere as a processing "island."

Sometimes, those expensive general purpose machines advertised as being able to do just about anything end up doing nothing well. JIT production has no use for machines like these. Instead, we should try to use only compact

Straight-line flow production

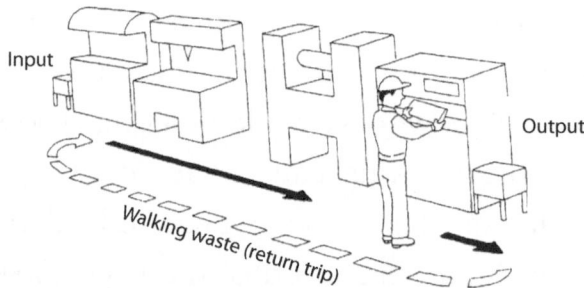

U-shaped manufacturing cell flow production

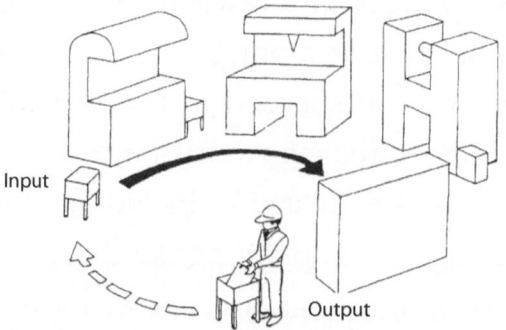

Figure 5.6   Flow Production Examples.

machines that can be arranged and rearranged into the line at a moment's notice and that are not so expensive that we have to worry about their capacity utilization rates.

Condition 8: Create U-Shaped Manufacturing Cells

This is another topic that does not directly relate to one-piece flow production. In some cases, it is fine to have a straight line for flow production. However, if we have one-piece flow production using multi-process operations, it is wasteful to require a worker who operates a series of processes along a straight line to walk all the way back from the final process to the starting one to get the next workpiece. This is where U-shaped manufacturing cells come in. (See Figure 5.6.)

What Is the Best Way to Eliminate This Kind of Waste?

We should try to arrange the input and output points as close together as possible. For short, we call this the "I/O matching principle." The closer the input and output points are, the less walking waste we will create.

These curved lines are called U-shaped manufacturing cells because they usually end up having a shape like the letter "U." However, they can just as well be arranged like circles or triangles if that works better. The exact shape of the cell should be determined based on such factors as the overall flow of goods in production, elimination of waste, and available space.

Of the above eight conditions, the most important by far is the first: one-piece flow. If we think switching to one-piece flow is too difficult and give up on it, we may end up handling lots of ten workpieces without ever realizing how much waste those breadbox-size lots create. People will start assuming that ten-unit lots are the smallest lot size possible in flow production.

But if we hang in there and manage to establish one-piece flow, we will hold the key to great success in eliminating waste.

The other seven conditions are like walls that protect the fortress of one-piece flow. Among these, Condition 4 (multi-process operations) would take prominence as the front wall and Condition 2 (lay out equipment according to the sequence of processes) would form the rear wall.

We can group these eight conditions according to the production factors they relate to most directly.

1. Equipment
   a. Condition 7: Make equipment compact
2. Equipment layout
   a. Condition 2: Lay out equipment according to the sequence of processes
   b. Condition 8: Create U-shaped manufacturing cells
3. Operation methods
   a. Condition 1: One-piece flow
   b. Condition 3: Synchronization
   c. Condition 4: Multi-process operations
   d. Condition 6: Standing while working
4. Operators
   a. Condition 5: Training of multi-process workers

Let it be clear from the outset that we can expect to run into many obstacles—equipment problems, capacity imbalances, and the like—as we work to establish these eight conditions in factory workshops. But the biggest obstacle is human resistance. We have to get people to drop all those tired old ideas, such as "This equipment can't be moved," or, "We'll lose money if we don't have lot production."

The best way to ensure success in establishing these eight conditions for one-piece flow production is to first get the people to "go with the flow" of JIT production.

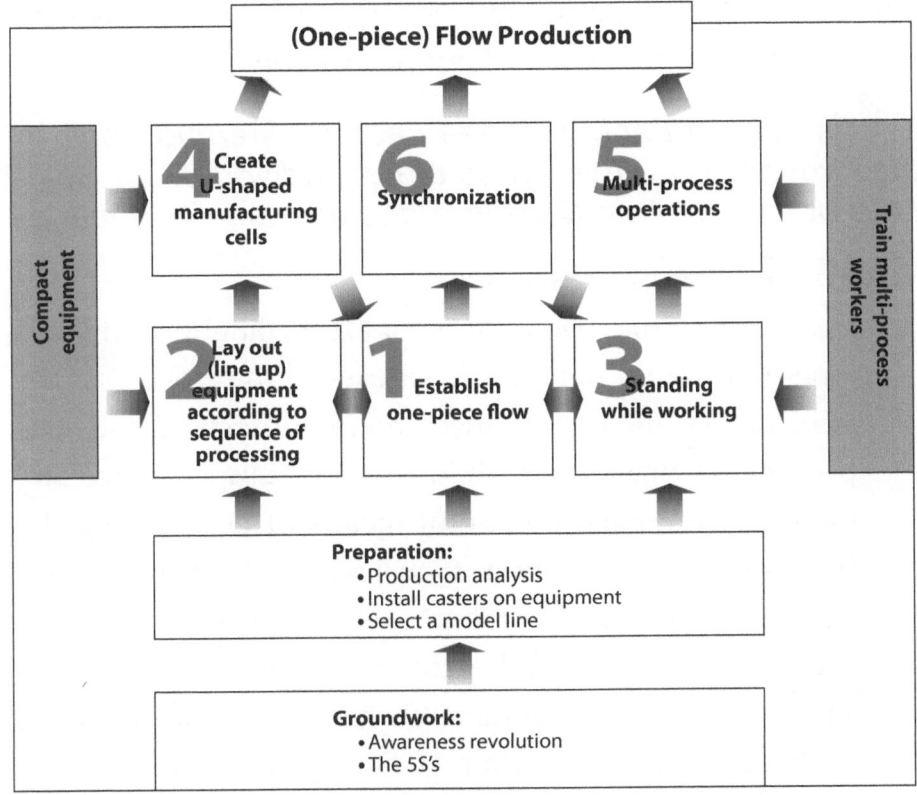

Figure 5.7   Interrelationship of Factors in (One-Piece) Flow Production.

Figure 5.8   In-House Seminar on JIT Production.

Steps in Introducing Flow Production

In establishing flow production—a basic part of JIT production—we need to rearrange the production equipment, but we do not have to find the perfect arrangement the first time. Instead, we should follow a series of experimental steps that well help us get closer to success. Figure 5.7 illustrates how various factors interrelate in flow production.

Let Us Look at Each of These Factors, Starting from the Groundwork—Two Types of Groundwork Must Be Laid before We Can Start Introducing Flow Production

1. The awareness revolution

Everyone at the company should be taught to discard long-established notions about everything from lot sizes to inventory and conveyance and to understand and support the JIT production philosophy. JIT study groups and in-house seminars are useful means of establishing the JIT awareness revolution. (See Figure 5.8.) (The awareness revolution is described in detail in Chapter 2.)

JIT production can be described and discussed in study groups and seminars. To really learn it, however, we have to practice it. After we have practiced the various procedures and steps for a while, we begin to develop a "feeling" for JIT; only then are we truly learning it in both heart and mind.

2. The 5S's

The 5S's are described fully in Chapter 4. The S's are the first letters in the Japanese words *seiri* (proper arrangement), *seiton* (orderliness), *seiso* (cleanliness), *seiketsu* (cleaned up), and *shitsuke* (discipline). The first two S's are the most important, and use two indispensable tools: the red tag strategy and the signboard strategy. All improvement activities should start with reinforcing the 5S's, particularly by using these two strategic tools.

Preparation for Flow Production

Once we have made some headway in establishing the awareness revolution and the 5S's, we are ready to enter the preparation stage for flow production. We can facilitate making improvements for flow production by analyzing the production data needed for building a model line, then selecting a model line.

	P-Q Analysis List					Analysis by: *J. Smith*	Date: *11/16/89*	

	P-Q Analysis List					Analysis Period: *10/1/89 to 10/31/89*		
No.	Item (part number)	Quantity	Total	%	Total %	Management category		
						A	B	C
1	RA1103	15,900	15,900	17.5	17.5			
2	RB0121	12,500	28,400	13.7	13.7			
3	RC1631	11,700	40,100	12.9	12.9			
4	RD1911	9,450	49,550	10.4	10.4			
5	RE0314	9,400	58,950	10.3	10.3			
6	RF1213	9,000	67,950	9.9	9.9			

Figure 5.9   P-Q Analysis List.

As a third preparatory step, we need to install casters on equipment units to facilitate their rearrangement into new layouts.

Preparatory Step 1: Production Analysis

Three types of analyses will help us understand flow production: P-Q analysis, arrow diagrams, and process path tables. We can use these three tools to eliminate waste and pave the way for lining up equipment according to the processing sequence.

P-Q analysis. The P stands for products and the Q for quantity (production output). By analyzing the relation between products and quantity, we can make a distinction between "flow of quantity" and "flow of product models." This will help us line up processes for flow production. The steps in P-Q analysis are described below:

Step 1: Obtain three or six months' data on product (or parts) and production output.

Step 2: Figure the total production output from the obtained data, list products in order of highest quantity to lowest quantity, then find their proportionate percentages. Write these on a P-Q analysis list, such as the one shown in Figure 5.9.

Step 3: Create a P-Q analysis table based on the P-Q list. (See Figure 5.10.) The vertical axis on this table indicates

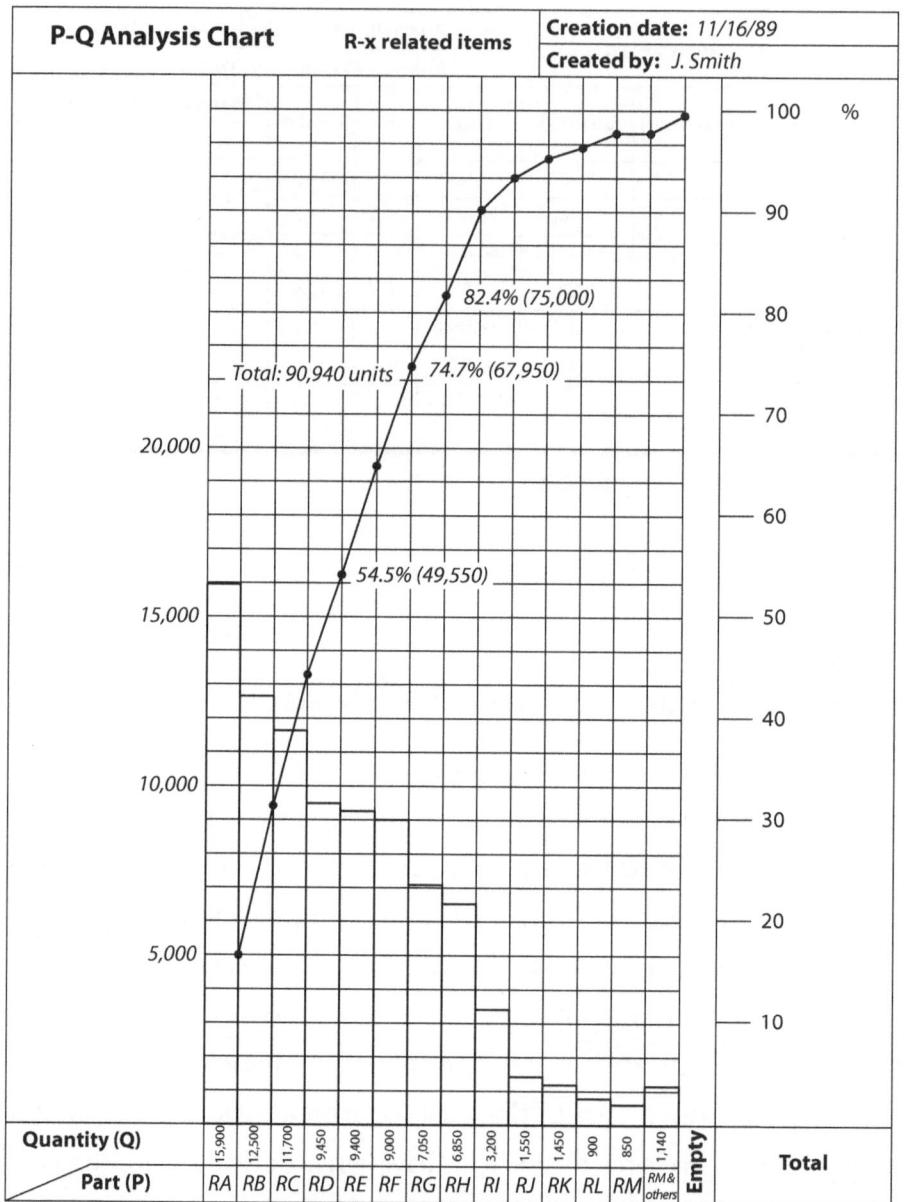

P-Q Analysis Chart			R-x related items			Creation date: *11/16/89*	

Total: 90,940 units

82.4% (75,000)
74.7% (67,950)
54.5% (49,550)

Created by: *J. Smith*

Quantity (Q)	15,900	12,500	11,700	9,450	9,400	9,000	7,050	6,850	3,200	1,550	1,450	900	850	1,140	Empty	Total
Part (P)	RA	RB	RC	RD	RE	RF	RG	RH	RI	RJ	RK	RL	RM	RM & others		

Figure 5.10   P-Q Analysis Table.

the production output (quantity) and the horizontal axis shows the products. Then we can use the output amounts to make an analysis of product groups A, B, and C.

Step 4: Design a line of processes based on the P-Q analysis list. As shown in Figure 5.11, the A group is a specialized line for building quantity, while the B group and C group lines are ordinary lines that build product models.

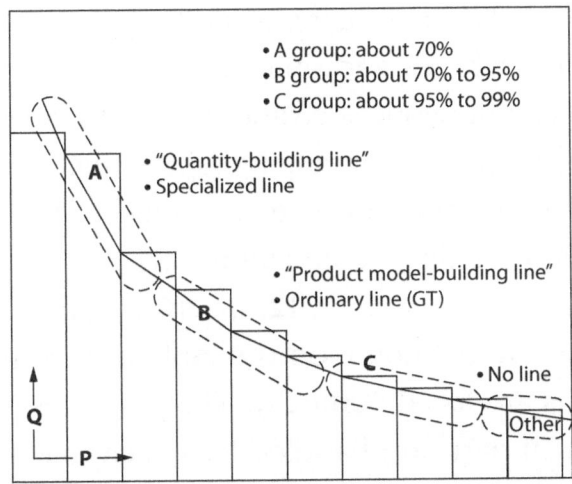

Figure 5.11   Line Design Using P-Q Analysis.

Process Path Table		Factory: *Tokai Plant*			Entered by: *Shin'ichi Yamagawa*		Date: *January 10, 1989*				
	Process name	Cutting	Drilling	Punching	Punching	Press	Press	Press	Bending	Bending	Bending			
No.	Item / Machine no.	M1	M2	M3	M4	M5	M6	M7	M8	M9	M10			
1	110931 (side board)	①		②		③	④		⑤		⑥			
2	130106 (side board)	①		②		③		④		⑤				
3	161137 (side board)		①	③	②		④	⑤			⑥			
4	1316171 (top board)		①	②		③			④					
5	1315021 (top board)		①		②			③		④				

Figure 5.12   Process Route Table.

The key factors in ordinary lines are Group Technology (GT) lines and changeover. GT lines are lines that turn out different products (or parts) that have similar process paths and can therefore use the same line configuration. We group such lines together as one line in the process path tables. We can improve GT lines by combining tool functions into fewer tools and by establishing simple changeover procedures.

Arrow diagrams. Before establishing flow production in the factory, we need to clarify how goods will flow and remove major forms of waste from retention and conveyance points. Arrow diagrams are tools for doing just these things. (Arrow diagrams are described in Chapter 3.)

Process route tables. Process route tables enable us to see what kind of machines and other equipment are needed for processing a certain workpiece and what path these processes should take. As such, they are indispensable aids for creating ordinary lines and grouping workpieces. These grouped lines are called GT lines. (See Figure 5.12.)

As can be seen in Figure 5.12, machines and other equipment are listed horizontally on the table and names of parts or other items are listed vertically. This provides a clear indication of which parts are handled by which machines and in which order. Once we can see this, we can more easily find the parts that use the same or similar machines in the same or similar order and group those parts together in a GT line. The main purpose of this type of GT line is to eliminate or greatly simplify the changeovers needed when switching to new product models.

Preparatory Step 2: Select a Model Line

Start this step by finding the most enthusiastic workshop in the factory, then make that workshop the model line. You can choose the model line based on the workshops involved in making a certain product, or based on specific processes or workshops. The important thing is to establish a model that clearly shows to everyone in the company how flow production works in a line and what kinds of things it involves.

The first thing to inquire about when selecting a model is the enthusiasm of workshop-level leaders, such as the foremen. Workshops that have weak leadership are much more likely to fail than those with strong leadership. Strong, energetic workshop leaders are a good sign of a highly active workshop.

Once you select a model line, put up a large sign with the words "JIT Model Line" and the target date for completion of the line. This will help cultivate the seeds of awareness and generate enthusiasm among the workshop staff for being chosen as leading examples for their factory. It will also help draw attention to what is happening in the model line.

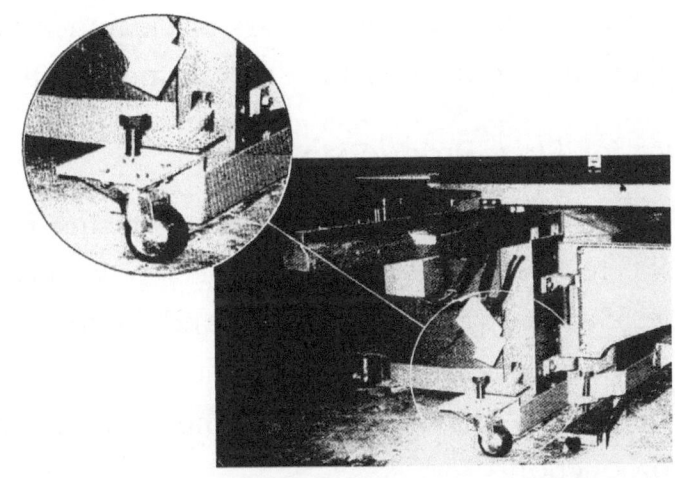

Figure 5.13   The Caster Strategy.

Preparatory Step 3: The Caster Strategy

It has always been a good idea to make equipment as movable as possible so that it can be easily rearranged into the most efficient layout for the particular product model being manufactured. However, many equipment units are bolted to the floor, the usual reasons being that they are either too bulky and heavy to move, or their high-precision mechanisms are too fragile to be moved. Once equipment gets bolted down, we must move the workpieces to the equipment rather than vice-versa. This makes one-piece flow production too difficult, encouraging factories instead to opt for shish-kabob production. Bolted-down equipment can make layout improvements difficult indeed. We need to put casters on as many equipment units as possible, so that we can rearrange machines, work tables, and other equipment whenever the need arises. In JIT, this is called the "Caster Strategy."

A word of caution about the caster strategy: Be sure to install casters on machines and work tables in such a way that they do not significantly change the height of the equipment. The photo in Figure 5.13 shows a "caster dolly" device that avoids having to install casters directly underneath the equipment. This device raises the equipment's height only slightly.

There should be about 10 millimeters of clearance between the frame and the floor to ensure smooth movability.

Procedure for Flow Production

We have finished the preparation for introducing flow production: We have launched the awareness revolution establishing the 5S's, and put various tools and strategies to use, such as production analysis, model line selection, and the caster strategy. Now it is time to follow the steps for introducing flow production.

Introductory Step 1: Use One-Piece Flow to Flush Out Waste

Flow production has two stages. The first stage comes before establishing JIT production and is concerned primarily with using one-piece flow to reveal concealed waste in the factory. The second stage is where we must establish the various conditions needed for full-fledged flow production, in which one-piece flow can be maintained without creating waste. Let us have a closer look at each of these stages.

■ **Stage 1: Revealing concealed waste with one-piece flow.**

At this stage, we need to "force" one-piece flow onto the current set-up, which means the current equipment, layout, and operation methods. This can be for just two processes, if you wish. Even if the workshop staff is reluctant and uncooperative, this "experiment" in one-piece flow production must be carried out.

At this point, it is best if we can train single workers to handle all of the processes that have been switched over to one-piece flow, but it can be done with a worker at each process, if necessary. It does not matter how odd or unorganized things look: Just carry out one-piece flow under the current conditions. This alone will flush out waste related to conveyance, large equipment, and unbalanced operations.

When waste has been revealed in this way, we confirm the waste and then eliminate it. This should not cost money. All we need is our wits and our muscles. This is what making improvements is all about.

This experimental switchover to one-piece flow for flushing out waste is also very important as a vehicle for teaching the spirit of JIT right from the start, before people have come to understand JIT fully. In other words, they are learning the form first to get a feeling for JIT. In this way, JIT improvement is an art similar to the oriental martial and aesthetic arts, such as karate, judo, flower arrangement, and the tea ceremony.

Figure 5.14 shows two diagrams of a diecast deburring line. This line includes two processes—a pressing process and a drilling process, each in a different workshop. The current set-up is for lot production; workpieces are handled in 500-unit lots.

Under this lot production set-up, no one notices the waste involved in conveying 500-unit lots along a distance of 120 meters. However, when we switch this over to flow production, each individual unit must be conveyed the 120 meters, and the waste becomes quite obvious. Once everyone has been impressed by how much concealed waste there was in conveyance alone, we can make an improvement to eliminate that waste. Obviously, this first switch to one-piece flow will mean considerably lower productivity. But making improvements involves more than simply raising productivity. Lowering productivity by revealing waste is a "teaching tool" that enables us to clearly recognize the waste.

■ **Stage 2: Maintain one-piece flow so as not to create waste.**

Once we have understood where waste lies in our conveyance system and operational imbalances, we

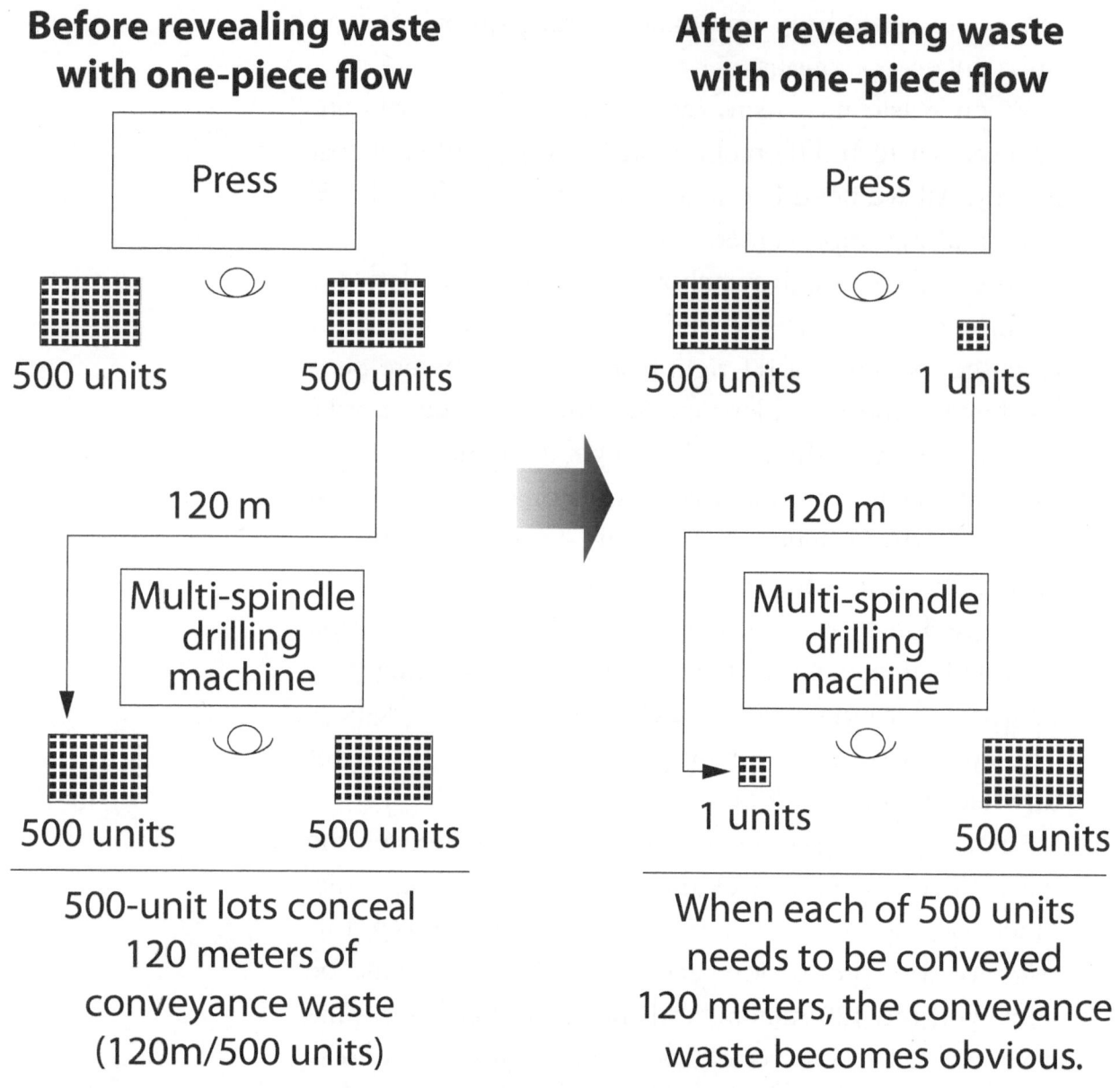

Before revealing waste with one-piece flow

Press

500 units   500 units

120 m

Multi-spindle drilling machine

500 units   500 units

500-unit lots conceal 120 meters of conveyance waste (120m/500 units)

After revealing waste with one-piece flow

Press

500 units   1 units

120 m

Multi-spindle drilling machine

1 units   500 units

When each of 500 units needs to be conveyed 120 meters, the conveyance waste becomes obvious.

Figure 5.14   Using One-Piece Flow to Reveal Waste.

can change the equipment layout into a closely-linked one-piece flow line to prevent this waste from being created again.

Figure 5.15 shows a line of cutting processes for automotive parts. Before making improvements, this line included widely separated workshops, was operated by four workers, and had multi-process operations only for some of the cutting processes.

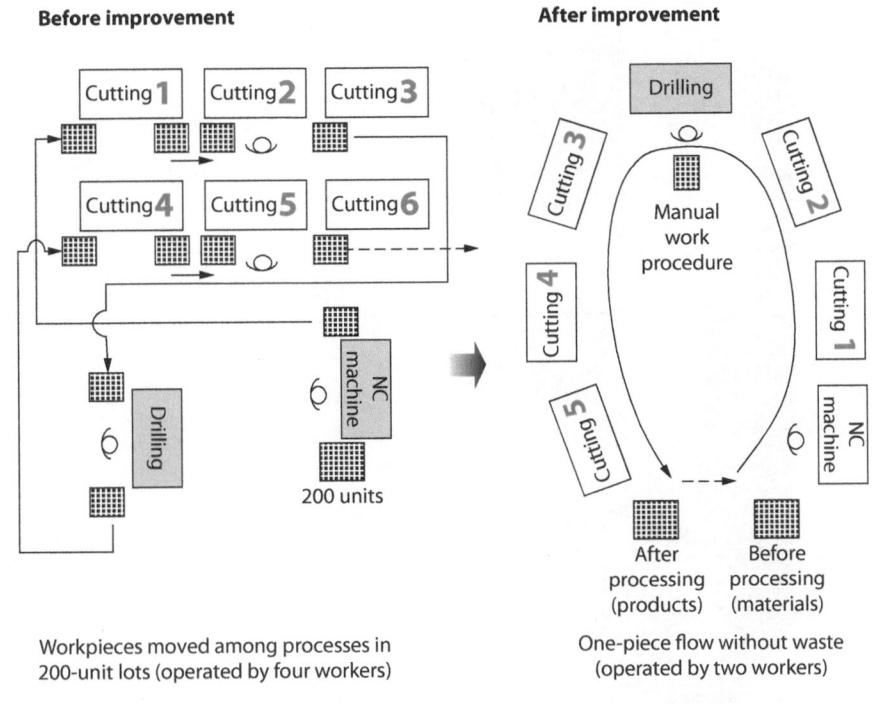

Before improvement

Cutting 1 Cutting 2 Cutting 3

Cutting 4 Cutting 5 Cutting 6

Drilling

NC machine

200 units

Workpieces moved among processes in
200-unit lots (operated by four workers)

After improvement

Drilling

Cutting 3 Cutting 2

Cutting 4 Manual work procedure Cutting 1

Cutting 5 NC machine

After processing (products) Before processing (materials)

One-piece flow without waste
(operated by two workers)

Figure 5.15 **Maintaining One-Piece Flow without Creating Waste.**

Then came the improvements. The scattered equipment units were brought together into a flow-oriented line from start to finish, and everything was set-up for one-piece flow production. This enabled the total elimination of in-process inventory, made the overall flow clearly visible and comprehensible to everyone, and enabled early detection of defects. Moreover, human work was separated from machine work, and this enabled a manpower reduction from four workers to just two.

Introductory Step 2: Arrange the Equipment in the Order of Processing

So far, we have pointed out conveyance waste, eliminated the conveyance system, set-up a way to move workpieces with a minimum of material handling, and rearranged the equipment layout. At this point, we are still faced with several problems. Many equipment units do not have casters and are difficult to move. And some of the larger equipment units are too big to fit directly into the line, which creates bottlenecks

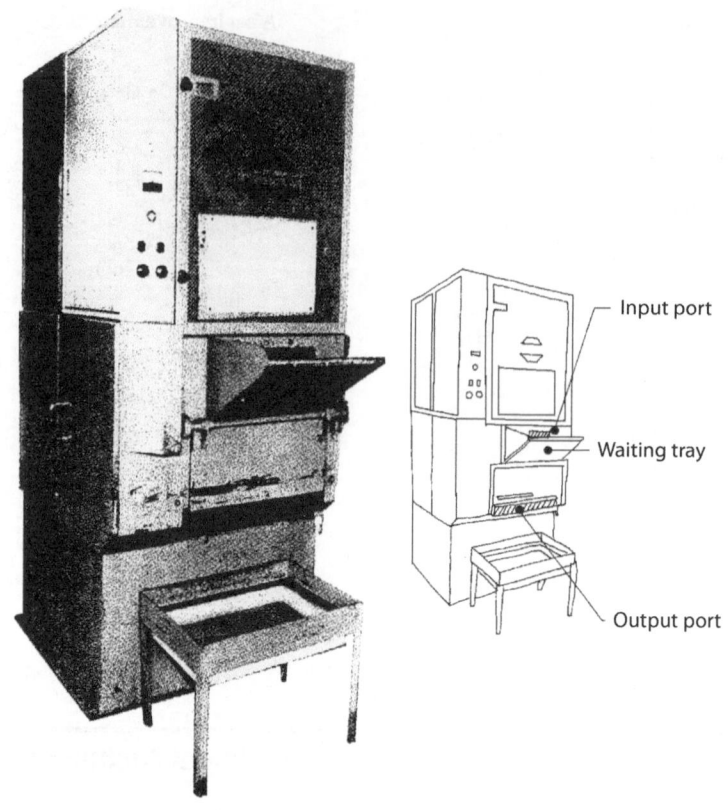

Figure 5.16 Compact Shotblaster.

at the isolated equipment units. Therefore, our next step is to find ways to make the equipment more movable and compact. Again, we should be using our wits and not spending much money to make these improvements.

There is no need to use large and expensive general purpose machines that can quickly process various kinds of workpieces in large lots. Instead, we need to be only as fast as the cycle time, and we must stress the need for compact, inexpensive machines that specialize in reliable, high quality processing of certain types of workpieces. To do this, we must develop skills in grasping the basic function of each process and selecting or designing equipment that serves precisely that function.

Figure 5.16 shows a newly developed compact shotblaster. Previously, lots of 500 units each were divided into large batches and loaded onto pallet containers for shotblasting.

They used a large shotblaster which can handle large batches, but cannot prevent the diecast units from colliding and denting each other. About 10 percent of the units were rendered defective by this shotblaster. The shotblaster's batch processing also meant that there were large piles of in-process inventory on either side of the shotblaster.

To eliminate dent defects and in-process inventory while reducing manpower, this company worked with the equipment's manufacturer in developing a compact shotblaster that could fit into the flow-oriented line.

Figure 5.17 shows an example of a compact washing unit. This washer is used to wash flax from soldered motor parts on a motor assembly line. Prior to this improvement, the parts were conveyed to a larger washer. This became quite impractical under one-piece flow production, especially since the previous system used lots of 200 workpiece units. The company made this compact washer, which was able to be inserted into the conveyor line, and this rearrangement alone eliminated the conveyance waste, retention waste, and manpower waste created by the large washing unit.

Preparatory Step 3: Standing While Working

We have gathered two or three processes into a line and have left the operators on their stools to operate one process each using one-piece flow instead of shish-kabob lots. The seated operators can hand-pass the individual workpieces down the line. Once this set-up starts working smoothly, we are ready for the first step in multi-process operations: standing while working. The operators should first learn to handle one process at a time on their feet. Standing while working has different characteristics depending upon the type of line involved. Let us look at how standing while working can be established first for an assembly line and then for a processing line.

Standing while working at an assembly line—Most assembly operations use conveyors to produce an even

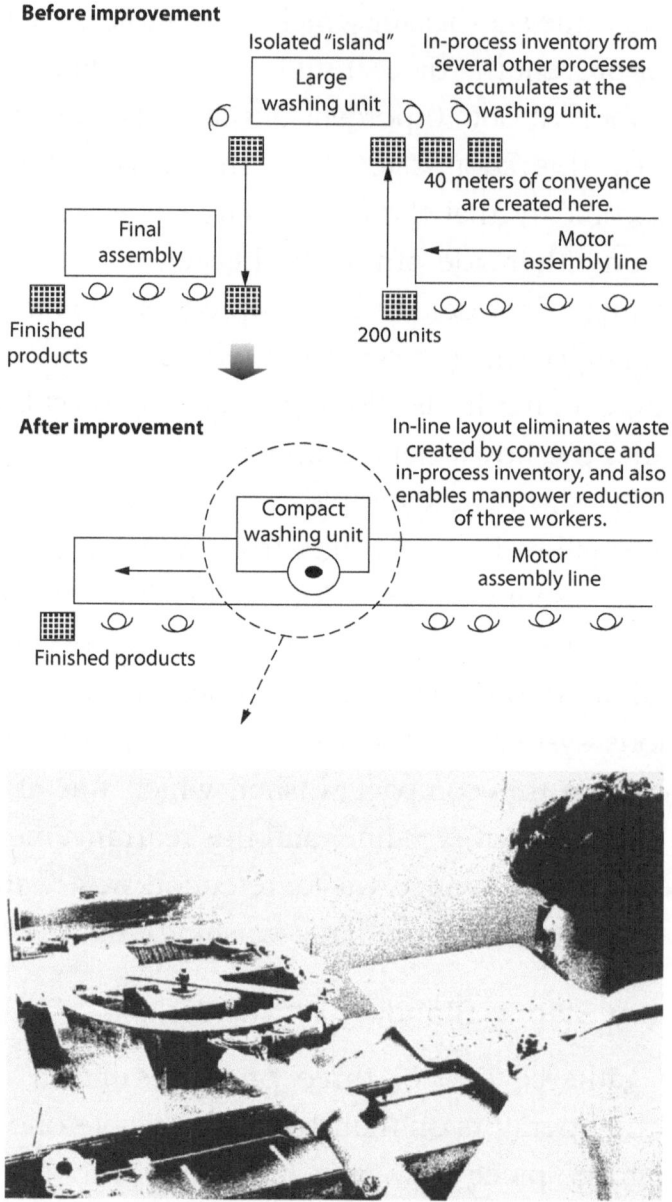

Before improvement

Isolated "island"

In-process inventory from several other processes accumulates at the washing unit.

Large washing unit

Final assembly

40 meters of conveyance are created here.

Motor assembly line

Finished products

200 units

After improvement

In-line layout eliminates waste created by conveyance and in-process inventory, and also enables manpower reduction of three workers.

Compact washing unit

Motor assembly line

Finished products

Figure 5.17 Compact Washing Unit That Fits into the Production Line.

production flow. Figure 5.18 shows workers standing while working at a VCR assembly line.

The photo in Figure 5.18 shows a free-floating assembly line that is 90 meters long. The workers in this photo had been seated while working until just a few days before the photo was taken. When the workers were seated, they tended to wait until workpieces were directly in front of them before

Figure 5.18   Standing While Working at an Assembly Line.

they picked them up to assemble them. Because the assembly workers were not balanced well (that is, they worked at different speeds), some workers spent a lot of time just waiting for the next workpiece to arrive.

Figure 5.19 shows a line balance analysis table that we can use to record the operation times for each worker. This table helps us understand how to rearrange labor at bottleneck-prone processes and achieve an overall balance in line operations.

However, such "analytical line balancing" does not always work well when put into practice. There are three main reasons why this can happen.

Reason 1: Rapid product diversification prompted the factory to switch product models while operation time analysis was still in progress.

Reason 2: At long last, we have finished the analysis. But by the time we are ready to put the results into practice, the corresponding product's life cycle has ended and the factory has switched to a new product.

Reason 3: The workers are part-time workers (such as working mothers) and the turnover rate is high. Absenteeism is also rather unpredictable.

Line Balancing Analysis Chart

Product name: PCB-01-03	Line name: Line A	COV.S s/m	By: Yamagawa	Date: November 16, 1988
Lot size: 300	Units per day: 300	ST (m)/unit	Time pitch: $\dfrac{\text{Total processing time}}{\text{Number of workers}}$	
Operating time: 480 seconds	Conveyor No.:	Item processing interval:		
Line balance loss = 100 − line balance efficiency = 32%		Line balance efficiency $=\dfrac{\text{Total of worker operations times}}{\text{Pitch time} \times \text{number of workers}}$	Conveyance workers: 8	
			Relief workers:	

Pitch = 96 seconds

Total pitch = 768 seconds

Net time: 523 seconds

Process times

Time	60'	72	56	75	82	57	69	52																					
Process name	Insert 1	Insert 2	Insert 3	Insert 4	Inspection	Soldering	Inspection	Assembly																					
NO	1	2	3	4	5	6	7	8	9	10	11	12	13	14	15	16	17	18	19	20	21	22	23	24	25	26	27	28	29

Figure 5.19   Line Balancing Analysis Table.

In the old days of high-volume lot production, product life cycles were longer, which made analytical line balancing a handy tool. In today's fast-paced world, there is not always time for this slow, analytical approach.

There are two alternative methods to analytical line balancing.

Method 1: "Practical line balancing." Here, we do not carry out any kind of analysis but instead simply start the product assembly operations, then take an ad hoc approach to changing the configuration of assembly workers whenever the need arises. This approach has two common names: "practical line balancing" and "the SOS system."

Specifically, we begin this approach by running the assembly line at a relatively slow pitch. Then we gradually accelerate the pitch until assembly workers who are not able to keep up sound an "SOS" alarm. The workshop leader then responds immediately by making a balance-improving adjustment to the assembly worker configuration.

This goes on repeatedly until the workshop members finally arrive at the best pitch and configuration for that particular product. At this point, things go much easier if the line uses forced conveyor rather than a free flow conveyor.

Method 2: "Baton passing zone method." Other names for this system are the "nonbalancing system" or the "cooperative system." This system avoids line balancing altogether.

In conventional conveyor operations, each worker is assigned a predetermined and fixed workload. This rigidity in worker responsibilities helps give rise to imbalances.

By contrast, the baton passing zone method gives each worker at each process a set of basic tasks to perform, as well as a set of overlapping tasks that are shared with the previous and/or next process. When each worker is finished, he or she can "pass the baton" to the person at the next process.

To recapitulate, the traditional "defensive" or "reactive" type of assembly operations, in which workers sat to work and held rigidly defined job duties, no longer works as well in today's manufacturing world. Instead, we need more "offensive" or "proactive" operations in which operators do more on their own to balance operations and ensure progress on the line. The latter type of operation is all the more necessary in view of today's ongoing trends toward production diversification, shorter product lives, and more and more part-time workers.

Standing while working at processing lines—Standing while working is much more common at processing lines than at assembly lines, and today almost all factories have processing line workers standing to work.

If anyone wonders why processing workers must stand while working, the answer is simple: They need to stand for multi-process operations. Standing should not be required just because it suits the conveyance system or because the equipment is too large to operate while sitting. When processing

Figure 5.20 Work Table Raised by Concrete Blocks for Standing While Working.

workers sit while working, they are like isolated little islands. We have to connect these little islands into an integrated line that follows the sequence of processes, and get workers to stand while working to enable one-piece flow and help build quality into products at each process.

Figure 5.20 shows how concrete blocks were used in one factory to raise the level of the work table to comfortably accommodate standing while working.

Preparatory Step 4: U-Shaped Manufacturing Cells

Flow production that includes two processes can be arranged in a straight line or an L-shaped line, as shown in Figure 5.21.

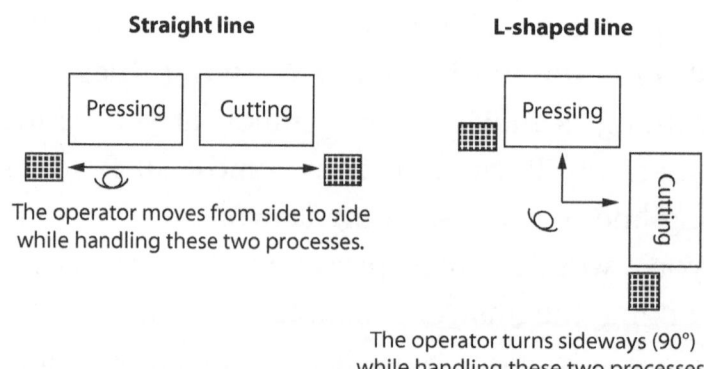

Figure 5.21 Two Types of Two-Process Flow Production Lines.

When the processing machines are small enough, they can be lined up side by side and the operator can move "crab-like" while operating both processes. If the machines are too large for this, they can be set at right angles to each other, and the operator can merely turn sideways to move the other machine. Either layout helps minimize the amount of "motion waste" created by the operator.

When there are three or more processes in one line, it is usually best to arrange them into a U-shaped layout. Although these sets of processes are called "lines" in Japanese, the name "cells" better conveys their function as a unit within the overall production line. We can minimize motion waste in these U-shaped cells by laying out the cell's input and output sites as close together as possible. Operators should always work on the inside of the cell, since this will enable them to get to each machine with fewer steps than if they were on the outside of the cell. It also makes it easier for teams of operators to help each other out whenever necessary. (See Figure 5.22.)

No matter what shape these cells take, the layout should work to minimize wasteful motions. Figure 5.23 shows how the layout does not have to be U-shaped, but instead can be

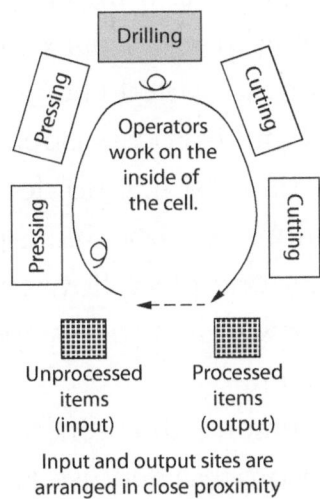

Figure 5.22   U-Shaped Cell.

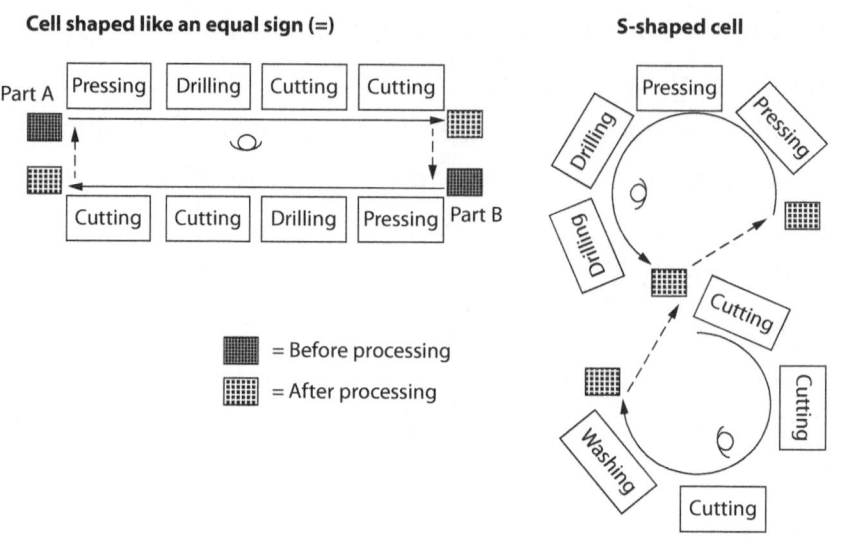

Figure 5.23 "Equal Sign" and "S-Shaped" Cells.

other shapes, such as a parallel line arrangement—like an "equal" symbol (=) or an "S" shape.

Preparatory Step 5: Multi-Process Operations

Once we give up the "one worker per machine" system and arrange the processing equipment according to the processing sequence, all sorts of new possibilities arise for worker operations. Most significantly, it lets us switch from single-process operations to multi-process operations. At first, the workers will have to get used to doing things a completely new way. Naturally, this will result in lower daily output for a while.

There must be no half-hearted changes. We cannot claim to have implemented multi-process operations if we are still handling workpieces in "shish-kabob" lots or "caravan-style." Multi-process operations is not multi-process operations unless it is done under one-piece flow conditions.

Figure 5.24 shows how multi-process operations were set-up for a sensor assembly line.

Before the improvement, this sensor assembly line had one sitting worker per process and used a conveyor to "push" finished lots toward the next group of processes. The manufacturing lead-time for products on this line was about

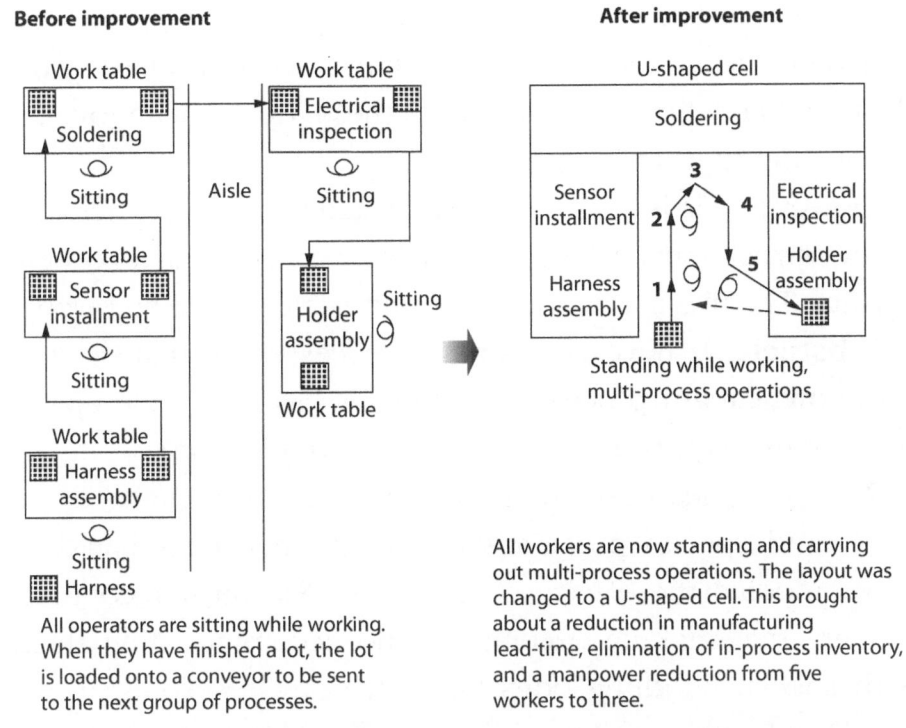

Before improvement

Work table

Soldering

Sitting

Aisle

Work table

Sensor installment

Sitting

Work table

Harness assembly

Sitting

Harness

Work table

Electrical inspection

Sitting

Holder assembly

Sitting

Work table

All operators are sitting while working. When they have finished a lot, the lot is loaded onto a conveyor to be sent to the next group of processes.

After improvement

U-shaped cell

Soldering

Sensor installment

Harness assembly

Electrical inspection

Holder assembly

Standing while working, multi-process operations

All workers are now standing and carrying out multi-process operations. The layout was changed to a U-shaped cell. This brought about a reduction in manufacturing lead-time, elimination of in-process inventory, and a manpower reduction from five workers to three.

Figure 5.24 Multi-Process Operations on a Sensor Assembly Line.

two days. The operators sat at work tables in cramped areas handling the lots that were passed to them at the upstream processes' convenience. These workers sat amid piles of in-process inventory.

After the improvement, the layout is a U-shaped cell, in which all workers are standing while working and handling multi-process operations under one-piece flow conditions. This arrangement reduced the lead-time and completely eliminated the in-process inventory. The cell only takes about a third as much space as it used to, and they were able to lower the cell's manpower requirement from five workers to three.

Preparatory Step 6: Synchronization

Synchronization means synchronizing both processes and workers so that the entire line and, eventually, the entire production system become synchronized. To do this, we must calculate the cycle time required for level production, after which we must match this up with the appropriate number of workers and the correct operational procedures. We must first build up a smooth rhythm within sections of the production line, then we can build these up into an overall production rhythm.

However, this is often much easier said than done. There are many obstacles that can stand in the way of achieving an overall rhythm. The five main types of obstacles are described below.

Obstacle 1: Several upstream processes bottleneck into one downstream process, resulting in inventory pile-ups at the downstream process. (*Solution: in-line layout.*)

Most factories have many "exceptions" to whatever rules exist, and special processes or procedures are created to accommodate these exceptional cases. We must recognize, though, that making exceptions and accommodating them with special handling does not solve any problems. *There need not be any exceptions in the factory.*

In many factories, people regard processes such as forging, casting, painting, washing, and calcination as "special processes." As a result, these processes get special treatment, and become self-involved little islands in the factory.

Figure 5.25 shows one such little island, a washing unit. Workpieces are conveyed from three cutting lines and piled up before this washing unit as in-process inventory. Before entering this washing unit, the workpieces are loaded by two workers into washing containers. Two other workers unload the containers and send the workpieces on their way downstream.

To solve this problem, we must remove the waste created by consolidating production flow at the large washing unit and then dispersing it again downstream. We can do this by incorporating small, inexpensive washers at the end of each processing line that formerly converged on the large washing unit. This in-line layout allows this factory to eliminate both the need for the four workers attached to the large washing unit and also the in-process inventory.

Obstacle 2: The "push" method, in which goods produced at one process are automatically sent to the next process, is

Before improvement

The washing process was set apart as a little island in the factory. Two workers were needed to load parts into washing containers and two others were needed to unload the containers and send the parts to the next processes. This arrangement still resulted in large piles of in-process inventory.

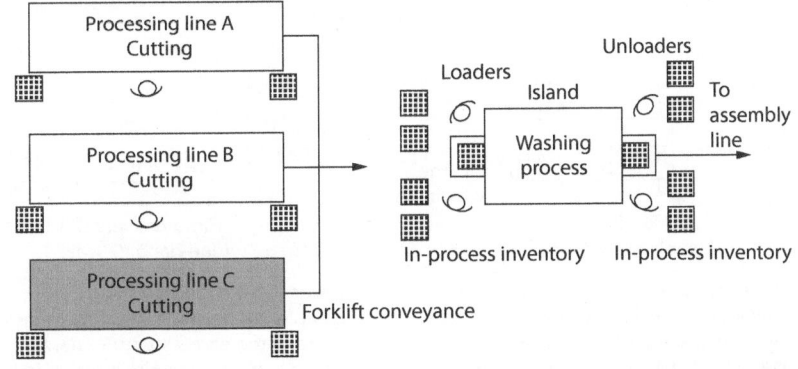

After improvement

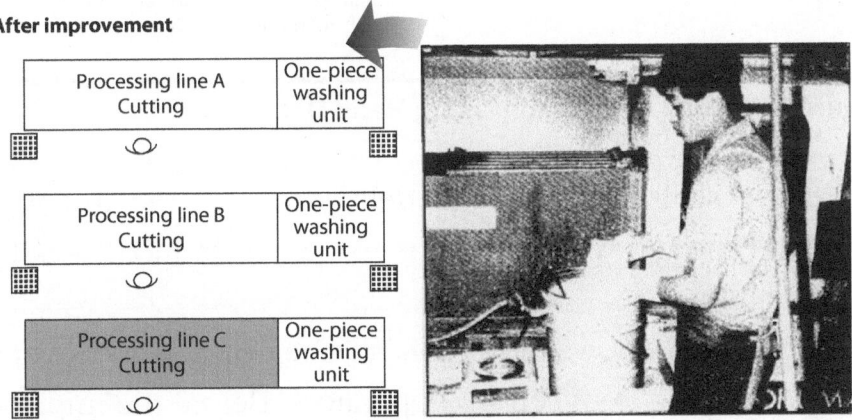

Retire the large washing unit and install small one-piece washers at the end of each processing line. This removes the need for a large washing unit and also eliminates the in-process inventory around the washing process.

Figure 5.25   In-Line Layout of Washing Units.

resulting in pile-ups of goods at certain downstream processes. (*Solution: the full work system.*)

The "push" method makes it hard to achieve a smooth flow of goods because automatically sending goods to the next process does not consider whether or not the next process is ready for the goods. The "pull" method is therefore highly recommended as a means to ensure a smooth flow of goods. We call the "push" method "the push system" and the "pull" method "the pull system."

There are various tools for implementing the pull system, such as *kanban,* hand delivering, and the full work system.

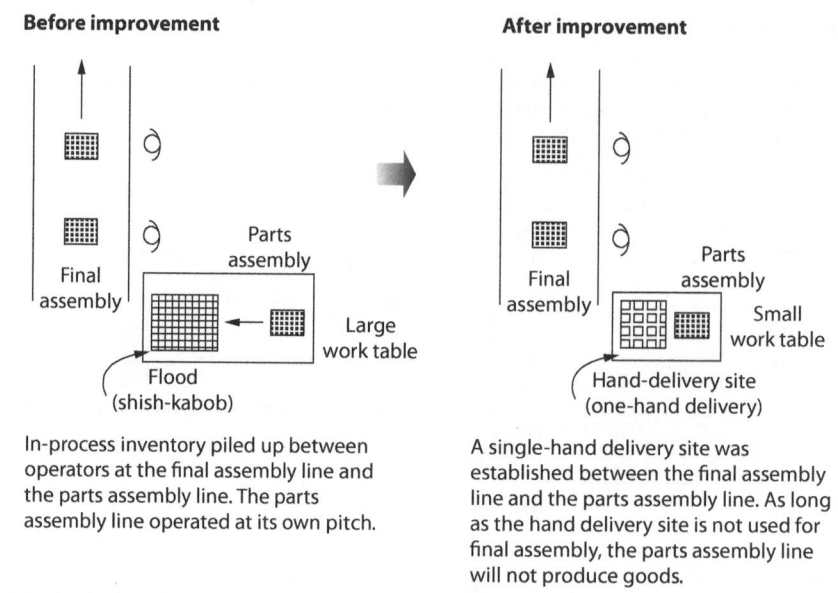

Before improvement

After improvement

Parts assembly

Final assembly

Large work table

Flood (shish-kabob)

Parts assembly

Final assembly

Small work table

Hand-delivery site (one-hand delivery)

In-process inventory piled up between operators at the final assembly line and the parts assembly line. The parts assembly line operated at its own pitch.

A single-hand delivery site was established between the final assembly line and the parts assembly line. As long as the hand delivery site is not used for final assembly, the parts assembly line will not produce goods.

Figure 5.26   Pull Production Using Hand Delivery.

Here, we shall look at hand delivering and the full work system. (The full work system is explained in more detail in Chapter 14.)

Figure 5.26 shows how the number of hand deliveries were calculated between two operators. Before the improvement, in-process inventory filled up the entire space between the final assembly and parts assembly lines and no one could find any way to synchronize the two lines. They responded instead by making the work tables smaller and reorganizing the physical space to make more room. They also set-up a place where goods could be hand-delivered, which meant there was one hand delivery. The improvement reduced all of the in-process inventory to this one hand-delivery. Furthermore, if the hand delivery can be eliminated, this improvement will enable implementation of the pull system and will make any imbalance between the final assembly line and the parts assembly line readily obvious. This improvement led to the following improvement.

Figure 5.27 illustrates the synchronization of a "pull system" involving a printed circuit board (PCB) assembly line and a DIP

Before improvement

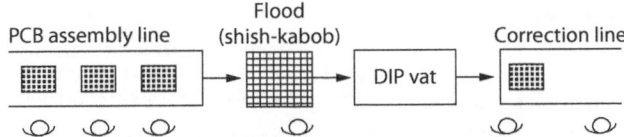

The PCB assembly line and DIP vat process did not operate at the same pitch, and this resulted in chronic accumulation of in-process inventory between them.

After improvement

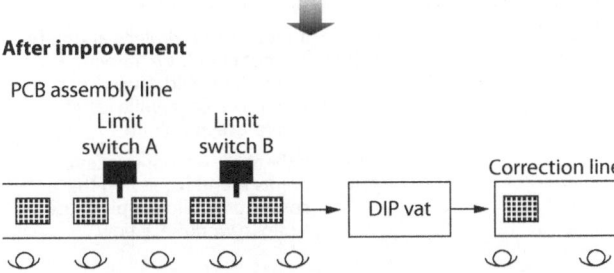

The conveyor for the PCB assembly was moved closer to the DIP vat, and use of two limit switches (A and B) enabled achievement of a pull production using a full work system. This eliminated the in-process inventory between the PCB assembly line and the DIP vat and enabled a reduction of one worker at the DIP vat process.

Figure 5.27   Pull Production Using a Full Work System.

vat. Before this improvement was made, the DIP vat process tended to lag behind, and there was a chronic accumulation of in-process inventory between the PCB assembly line and the DIP vat. The improvement included moving the two processes closer together and installing two limit switches (A and B) to enable implementation of a full work system. This improvement eliminated in-process inventory between the PCB assembly line and the DIP vat and led to manpower reduction on both the PCB assembly line and the DIP vat process.

A full work system controlling points A and B proved necessary for achieving pull production and synchronization with downstream processes, as shown in Figure 5.28.

Obstacle 3: Variation in work procedures among different workers causes delays or idle time. (*Solution: cooperative operations.*)

Whether it be a processing line or an assembly line, balanced operations among workers within the line is a key prerequisite for maintaining a smooth flow of goods. Such

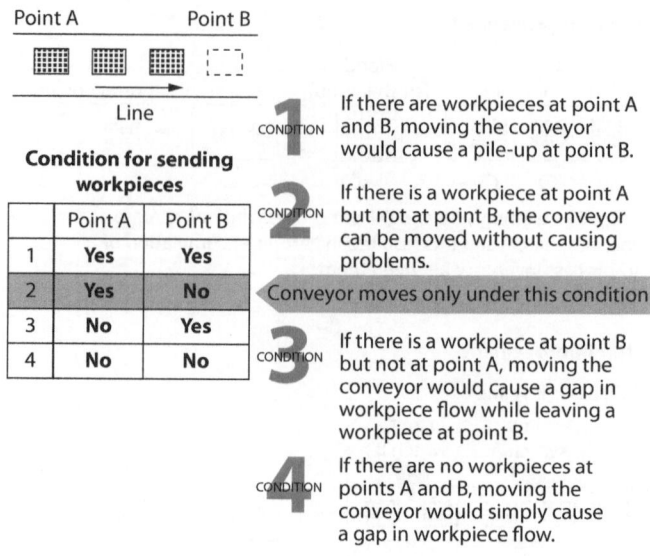

Figure 5.28   Devices Enabling a Full Work System.

balancing of operations takes a lot of training and practice, but these days more and more factory workers (in Japan) are part-time workers, which makes it harder to achieve and maintain such a balance. As a result, maintaining a constant tact time for operations such as fine-tuning electronic products is difficult indeed; delays or idle time often occur, upsetting the balanced flow of goods.

The solution that will keep balanced operations and smooth flow of goods from being upset is to have workers consider their operations flexible rather than rigid. In other words, they should be trained to help other workers when the flow starts becoming unbalanced. This approach is known as "cooperative operations" or the "baton passing method."

Figure 5.29 illustrates the steps to take in carrying out cooperative operations on an assembly line.

■ *Step 1: Standing while working.*
   This step starts with having all of the workers on the assembly line stand up. They should perform their operation whenever a workpiece arrives in front of them. This means abandoning their old "reactive" way of working

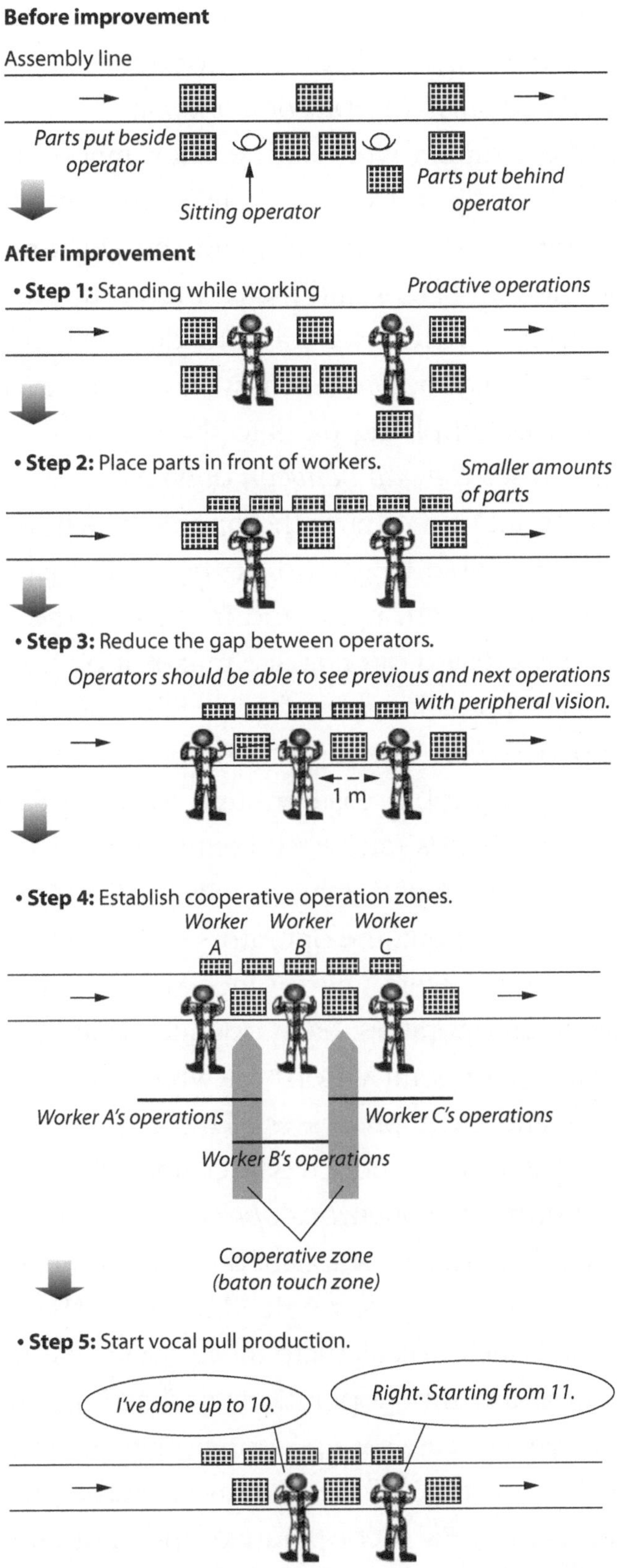

Before improvement

Assembly line

Parts put beside operator

Sitting operator

Parts put behind operator

After improvement

- **Step 1:** Standing while working   *Proactive operations*

- **Step 2:** Place parts in front of workers.

  Smaller amounts of parts

- **Step 3:** Reduce the gap between operators.

  Operators should be able to see previous and next operations with peripheral vision.

  1 m

- **Step 4:** Establish cooperative operation zones.

  Worker A   *Worker B*   *Worker C*

  Worker A's operations   *Worker C's operations*

  Worker B's operations

  Cooperative zone (baton touch zone)

- **Step 5:** Start vocal pull production.

  I've done up to 10.   *Right. Starting from 11.*

Figure 5.29   Improvement Steps for Cooperative Operation.

and adopting a "proactive" method that emphasizes the value each worker adds to the product.

■ *Step 2: Place parts in front of workers.*

When the workers were sitting, they made little use of the area directly in front of them. Standing while working enables workers to eliminate the stacking of parts on their left and right and instead have all parts in front of them. To do this, we have to decrease the amount of parts placed before each worker and increase the frequency of supplying parts to the workers.

■ *Step 3: Reduce the gap between operators.*

By placing all of the parts the operators will be using in front, we are able to get rid of the parts that had been piled up on the left, right, and in back of the operators. This newly created open space makes it obvious that the operators are too far apart from each other. In reducing the gap between operators, we should figure that the operators should be close enough to reach each other's outstretched hands (about 80 centimeters to 1 meter). In assembly operations for home electronics products and electrical equipment, the operators should be even closer; about 60 centimeters apart. Once we have reduced the gap between operators, each operator is able to keep an eye (using peripheral vision) on what is going on at the previous and next processes. This creates an environment that is more conducive to cooperative operations.

■ *Step 4: Establish cooperative operation zones.*

Now that we have established a layout that supports cooperative operations, we need to establish cooperative operation zones. To calculate these zones, we need to list each of the assembly operations and assign a number to each. Then we can set-up cooperative operation zones that can cover some of the operations at the previous and next processes. Each cooperative operation zone should be expressed as starting from one operation number and ending at another operation number, as in the cooperative

Cooperative Operation Zone Checklist		Factory: *Chiba*			Product: *PCB 1013*					
		Section: *1st Assembly Dpt., Line A*			By: *Yamagawa*			Date: *1/4/1989*		
Process No.		1	2	3	4	5	6	7	8		
No. \ Operator name / Parts input		Tucker	Engle	North	Brown	Meyer	Kline	Jones	Black		
1	11-1640-20	○									
2	16-1311-31	○									
3	19-2931-16	●	●								
4	20-2131-16	●	●								
5	14-1923-61		○								
6	36-3111-21		○								
7	63-1416-41		●	●							
8	27-2131-51		●	●							
9	32-8136-24			○							

Figure 5.30 Establishment of Cooperative Operation Zones.

operation checklist shown in Figure 5.30. The zones of cooperative operations among operators at adjacent processes is reminiscent of the zones on the running track within which relay runners must pass their batons. That is why cooperative operations are sometimes called the "baton passing method." In track relays and in cooperative zones on the assembly line, the "baton pass" can be made anywhere within the baton passing zone.

■ *Step 5: Start vocal pull production.*

In this case, "vocal pull production" means that the worker—who is "passing the baton" by turning the rest of the process's operations in a cooperative operation zone over to the next worker—should vocally confirm which operation number he or she has finished. This helps prevent any misunderstanding between workers that might result in the repetition or omission of an operation.

Obstacle 4: When we have shish-kabob production on the assembly line, it is not possible to synchronize the assembly line with the process line, which also means that the flow of goods cannot be synchronized. (*Solution: establish specialized lines.*)

Many factories have assembly lines that are used to put together a variety of product models. When asked why they do it this way, the managers of such lines always have some excuse, such as: "We don't have any other equipment," or "There's no room to do it otherwise," or "These are our most efficient workers."

When several different product models are assembled on the same line, the many equipment changeover operations are bound to be a haphazard affair, and the line will likely adopt shish-kabob production to minimize the number of required changeovers. This reinforces all the old conventional notions about manufacturing and creates a vicious cycle.

Figure 5.31 shows how a mixed-product assembly line can be changed into three specialized assembly lines. Before this improvement, one assembly line operated by ten workers would handle three product models per day. This resulted in a lot of waste created by changeovers and by unbalanced operations following each changeover. Also, because the line was using the shish-kabob production method, it was quite difficult to synchronize the assembly line with the processing

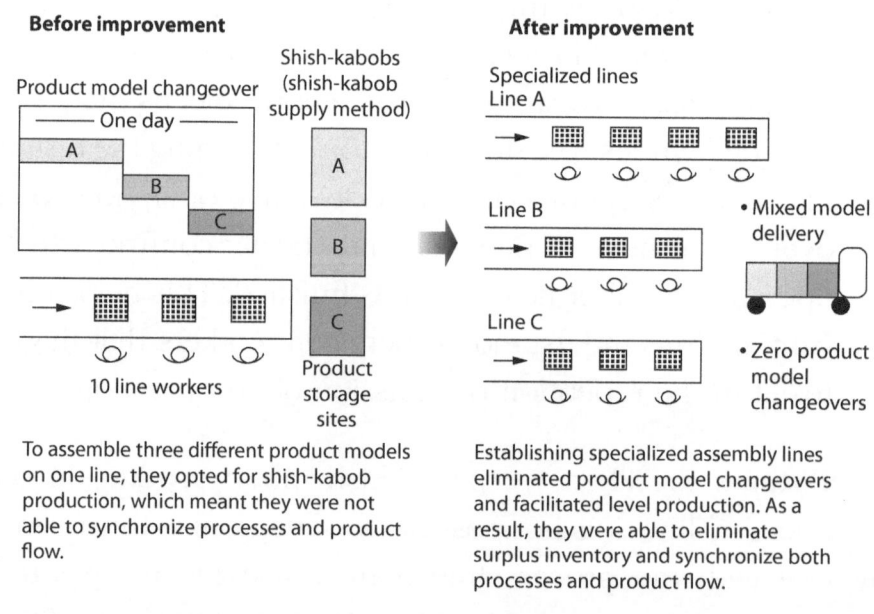

Figure 5.31   Improvement to Establish Specialized Assembly Lines.

line and to synchronize the flow of goods. This led to large amounts of inventory.

After the improvement, the workers were divided into the numbers needed for the required output of each of the three product models (A, B, and C) to enable level production. This also completely eliminated the need for changeovers, prevented disruption of balanced operations, and made for easier and smoother synchronization of the assembly and processing lines, and of the product flow. Finally, it enabled the elimination of surplus inventory.

Obstacle 5: Attempts to reduce the number of change-overs in the processing line results in large lots, which disrupts the smooth flow of goods. (*Solution: improve the changeover procedures.*)

When changeovers for different product models occur in the assembly line, they usually also take place in the pro-cessing line. To avoid the hassle of frequent changeovers, the lines naturally tend toward handling large lots, which disrupts the flow of goods and makes it hard to synchronize upstream and downstream processes.

We might think that the same advantages can be realized by also setting up specialized processing lines for different product models. However, unlike assembly lines, processing lines require various expensive types of machines. It is there-fore necessary to make each processing machine handle sev-eral different product models. In such cases, the appropriate improvement is to improve the changeover procedures. Changeover improvements are described in Chapter 11.

Case Study: Flow Production within the Factory—Improvement at a Diecast Factory for Automotive Electrical Parts

The factory in this case study, a subcontractor to an auto-mobile manufacturer, makes diecasts for automotive electrical parts. Before making improvements, this factory operated entirely on the shish-kabob production system, using lots of 500 to 700 units loaded into containers and conveyed between processes by forklift. The factory was operating slightly in the red, but the company somehow managed to balance its accounts at the end of each term. The factory managers decided to adopt JIT improvement as a way to revolutionize their tired old factory management system.

Before Improvement

Figure 5.32 shows this factory's processing sequence and production flow prior to improvement.

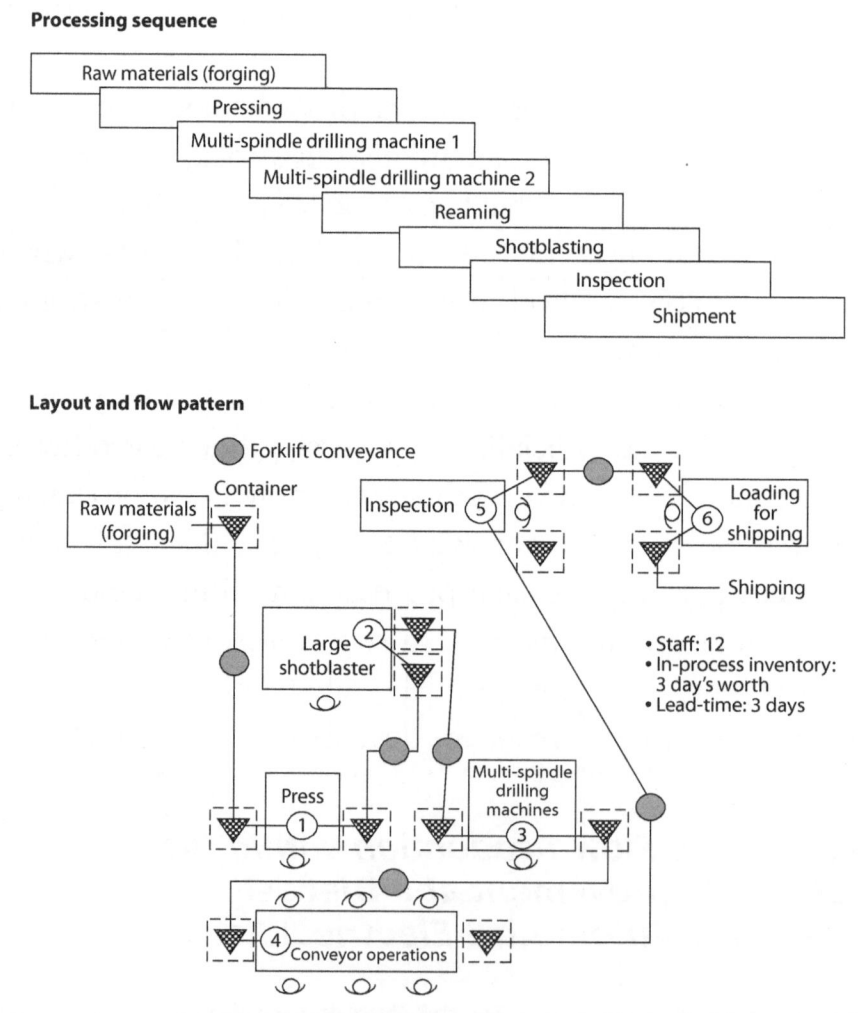

Figure 5.32   Processing Sequence and Production Flow for Diecast Product A (before Improvement).

The major characteristics of this factory are:

■ Layout: Job shop layout; similar tasks are grouped within the same workshops.
■ Production flow: Shish-kabob production using lots of 500 to 700 units.
■ Operators: Single-skilled workers, each assigned to a specific process.
■ Machines: Large machines capable of handling large lots.

Under this arrangement, it takes 12 workers to operate the line for product model A, and it takes three days for each workpiece to go all the way from the forging process to shipment. The factory contains three days of in-process inventory and the lots are conveyed between processes via forklifts requiring full-time forklift drivers.

The biggest obstacle to improvement was the large shotblaster, shown in Figure 5.33. Every workpiece that this factory handled had to be shotblasted by this big machine, and naturally this led to large piles of in-process inventory on the upstream and downstream sides of the shotblaster. In addition, the fact that workpieces were shotblasted in large batches meant that the workpieces got jostled around in the shotblaster. Inspectors were needed to sort the damaged diecasts from the undamaged ones.

Figure 5.33 Large Shotblaster.

Processing sequence

Raw materials (forging)

Pressing

Multi-spindle drilling machine 1

Multi-spindle drilling machine 2

Reaming

Shotblasting

Inspection

Shipment

Layout and flow pattern

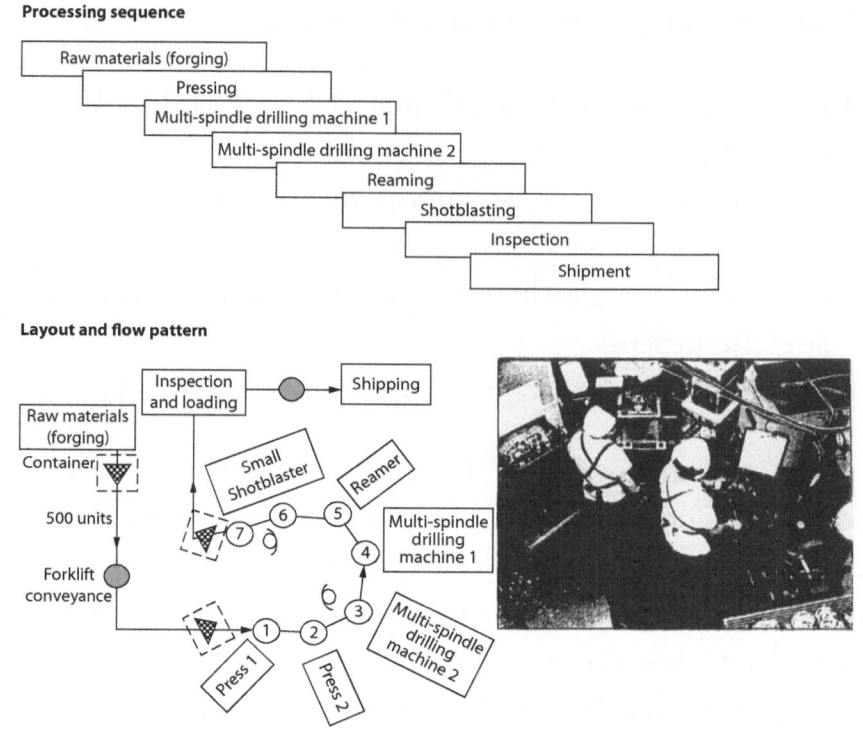

Figure 5.34   Processing Sequence and Production Flow for Diecast Product A (after Improvement).

After Improvement

We got right to work by selecting a model line for manufacturing product A. We abandoned the job shop layout and switched over to a flow shop (line) layout that emphasizes the flow of goods. At this point, we also abandoned all of the manual deburring processes and switched over to machine operations using a press, multi-spindle drilling machine, and other equipment. This enabled us to eliminate all manual processing.

Figure 5.34 shows the processing sequence and production flow following the improvement.

The major characteristics of this factory are:

■ Layout: Flow shop layout (in-line); emphasizes the flow of goods.

■ Production flow: Workpieces exit the forging process in 500-unit lots and move in one-piece flow from the pressing process to shipment.

Figure 5.35 Compact Shotblaster for In-Line Layout.

- Operators: Multi-skilled workers, trained to handle seven processes, from pressing to shipping.
- Machines: Eliminated large shotblaster and built a small shotblaster conducive to in-line arrangement. (See Figure 5.35.)

As a result of this first improvement, the model line was able to manufacture product model A using only two workers instead of 12. To reduce the former lead-time of three days, this improvement brought about a cycle time of 10 seconds for one-piece flow. Naturally, the inventory was also drastically reduced, reaching zero except for seven workpieces of inventory at the pressing processes and three at the drilling machines.

In addition, this improvement meant that forklift conveyance was no longer needed within the line. Furthermore, the elimination of the large shotblaster did away with the chronic problem of shotblast-damaged diecasts.

After their initial success with this model line, the factory managers extended the improvement laterally to other lines.

Within two years, the company's business accounts were in the black.

Flow Production between Factories

Applying the Flow Concept to Delivery

When we take a successful model of flow production, such as the model line described above, and extend that clearly visible example to other lines in the same factory, we call it "lateral development."

Once we have carried out lateral development and have established a firm footing for flow production within the factory, we are ready to take on the challenge of extending these improvements outside of the factory. In so doing, JIT production begins to take on greater height and depth as well as breadth.

Obviously, this vertical development of JIT improvements is centered on the factory where the improvements began and is generally extended in two directions: the "delivery" direction, which means from the factory to its vendors and subcontracted suppliers, and the "shipment" direction, which means from the factory to its customers or wholesalers. Once we understand these two directions, we must also understand that the most important direction is that between the vendor and/or subcontractors and the factory.

JIT's basic approach is to reduce the amount of each delivery and to compensate by increasing the frequency of deliveries. Obviously, if the deliveries are more frequent, they will also be more costly if current methods are used.

Let us suppose that deliveries are increased from once a day to twice a day and the per-delivery amount is correspondingly cut in half. This means the deliverer's cost will be approximately double.

When people hear this, many are quick to conclude that the JIT production system bullies the subcontractors. But this is not so. The general trend toward diversification and shorter delivery deadlines has affected the distribution industry, the manufacturing industry, and the transport industry. Right now, the transport industry is confronting this challenge. Meanwhile, manufacturers are struggling to meet market needs for product diversification and short delivery scheduling.

Several clever new delivery methods have been developed. These methods concern three main aspects of delivery operations: loading methods, frequency of delivery, and transport routes.

Loading Methods

The product diversification trend has radically changed loading methods. Cargo loads used to be mainly all the same type of products. Today we not only have mixed-product loads, but also mixed-product loads that are loaded in the sequence of their use on the client's production line. (See Figure 5.36.)

• **Single-product load**

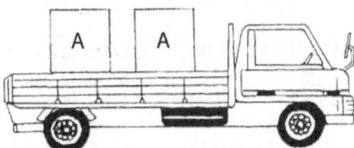

• **Mixed load**

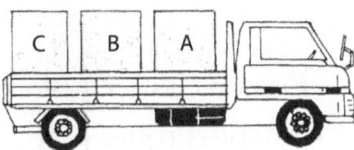

• **Sequential mixed load**

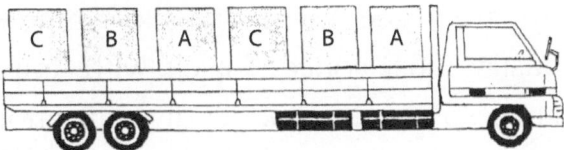

Effects of the product diversification trend

Figure 5.36   Loading Methods.

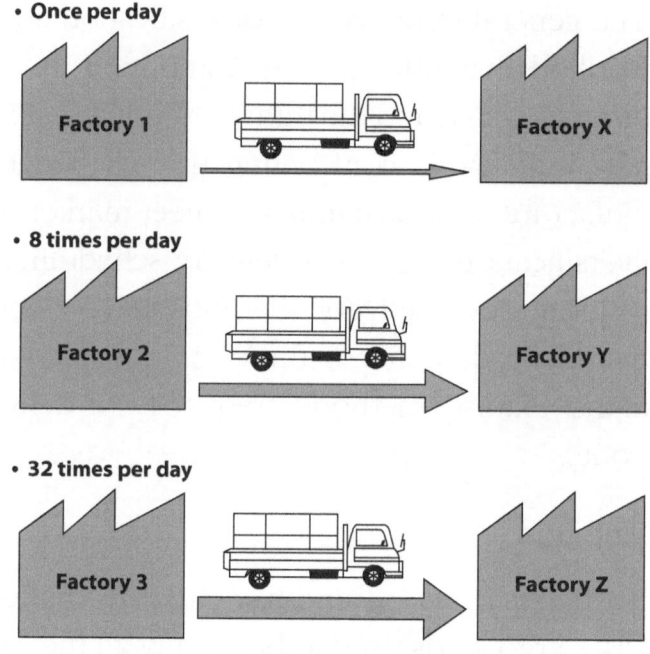

Response to need for less inventory and shorter delivery times

Figure 5.37 Frequency of Deliveries.

Frequency of Deliveries

Product diversification can easily lead to greater inventory. To keep inventory levels down and lead-times short, we must have more frequent deliveries. Sometimes we must switch from just one delivery per day to eight per day, from eight to 16, or even from 16 to 32. (See Figure 5.37.)

Transport Routes

One way to hold down the higher costs caused by product diversification is to improve transport route planning. Instead of simple point-to-point deliveries, it may be more economical to make circuit or compound deliveries. (See Figure 5.38.)

Thus, there are three main areas of improvement the transport industry must concern itself with: improved loading methods in response to product diversification, more frequent deliveries in response to lower inventory levels and shorter lead-times, and improved transport route planning in response to the need for cost reduction.

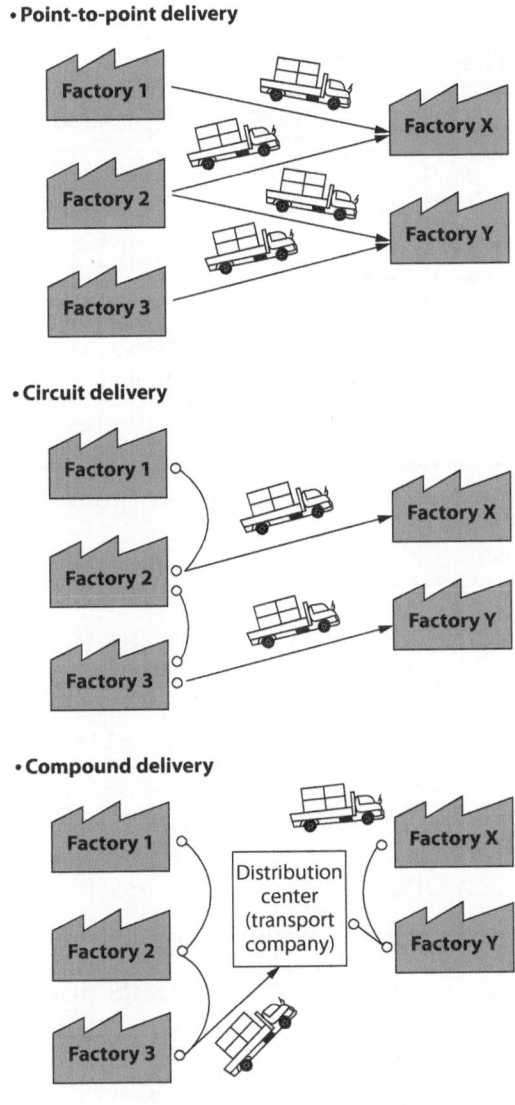

Response to cost reduction needs

Figure 5.38   Transport Routes.

It follows that the best combination of improvements is when the transport company manages to implement sequential loading, 32 deliveries per day, and compound deliveries.

The transport industry is witnessing a major shift away from large-scale container deliveries and toward smaller packages delivered door-to-door. In big cities, we can even find small package deliveries being made via motorcycle. Whenever there are new needs, the transport industry is obliged to respond with new methods.

Delivery Company Evaluation Chart				Factory: Tohoku Plant														
				Name/Dept. of evaluator: Anderson, Purchasing dept.									Date: November 16, 1988					
No.	Company	Main product	Manager (in-house)	Manager (delivery company)	Loading method			Frequency of deliveries							Transport route			Total
					1	2	3	1	2	3	4	5	6	7	1	2	3	
1	M Company	Resistors	Off	Jones	○			○							○			1
2	Y Company	A1 units	Lennon	Sandler		○				○					○			6
3	K Company	C materials	Lennon	McTighe		○			○						○			4
4	F Company	Packaging	Off	Rosen		○							○		○			10
5	T Company	Coils	Smith	Amick	○				○						○			2

Figure 5.39   Delivery Company Evaluation Table.

Figure 5.39 shows a delivery company evaluation table. Factory managers can use this table to evaluate how well each delivery company responds to their needs and to help improve their own factory's policy on deliveries.

Applying the Flow Concept to Delivery Sites

In JIT production, the secret for success in deliveries is not the conventional wisdom of delivering larger loads in fewer trips. It is just the opposite: smaller loads and more trips.

For instance, assuming there are 20 workdays in a month, consider the following two monthly delivery schedules:

A. Deliver once a month, 100 units per delivery (= 100 units total).

B. Deliver 20 times a month (daily), 5 units per delivery (= 100 units total).

In JIT production, we choose the latter. Even though the delivered units add up to the same total, the delivery methods are as different as night and day. Method B calls for 20 times more deliveries than Method A.

Next, we need to consider another very important issue: Which part of the factory should take in the delivered items?

Exactly where and how these deliveries are made can have a big impact on the handling of materials and parts in the factory.

The following are five points to remember for setting up delivery sites that will help prevent goods from accumulating and will make for a smooth flow of goods with little or no waste.

Point 1: Self-Management by Delivery Companies

In principle, the delivery company should be responsible for managing the delivery site it uses. In other words, the delivery company should bring the cargo all the way to the delivery site, keep the site properly arranged and orderly, and manage its general condition.

I strongly suggest that signboards be used to clearly indicate who brings what to where and exactly when. (See Figure 5.40.)

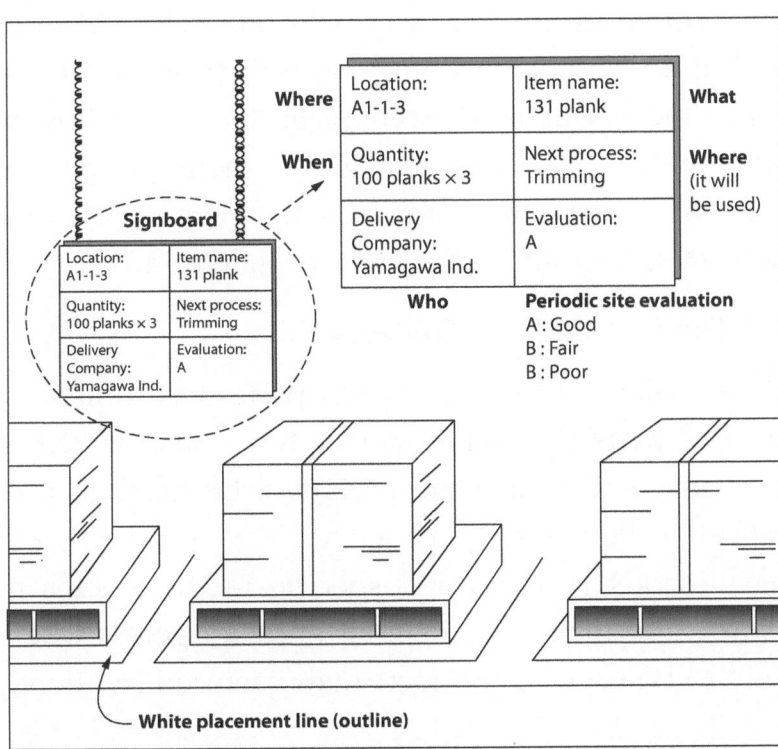

Figure 5.40 Establishment of Delivery Sites and Signboards for Delivery Site Management.

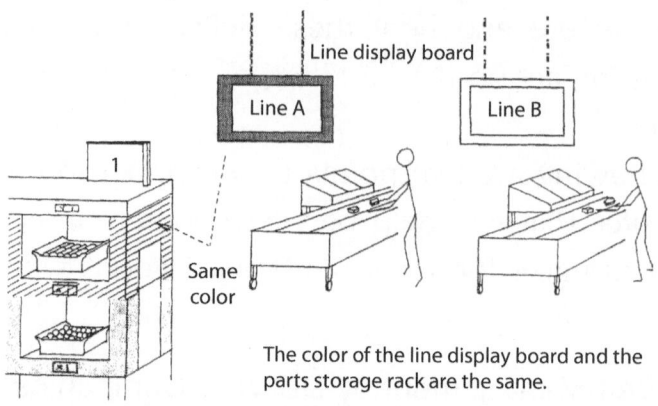

The color of the line display board and the parts storage rack are the same.

Figure 5.41   Line-Specific Method of Color Coding.

Point 2: Color Coding for Orderliness

Color coding is a good way to clearly show the relationship between things and processes—that is, what things are used in which processes. One good way to color code this relationship is to select a different color for each line and use that same color for the parts and materials that will be used in that line.

Color coding in this way will help prevent parts mix-ups when parts are supplied to the various lines at the factory. At the same time, it will also help parts and materials flow more smoothly to the lines with less waste, thereby contributing to an overall smooth flow of goods. (See Figure 5.41.) (Color coding is described in more detail in Chapter 4.)

Point 3: Product-Specific Delivery Sites

There are basically two ways to sort parts: according to similar types of parts that serve similar functions, or according to the products in which the parts will be used. These are respectively called "function-specific" and "product-specific" sorting methods. The product-specific method helps minimize waste and makes for a smooth flow of goods when the parts are to be used in products manufactured frequently.

Point 4: FIFO (First In First Out)

Whenever goods are put somewhere, there is always a process of placing and retrieving. If the most recently placed product is the one to be retrieved, we call it a LIFO (Last In First Out) arrangement. The problem with this arrangement is that the oldest item (the one placed there first) is also the last one to be retrieved. Delays in retrieving stored products can make these older items grow very old indeed.

Obviously, this is not a desirable situation. Therefore, we should be sure to have the opposite arrangement—FIFO (First In First Out)—whenever possible, to keep items moving as if they were on a conveyor belt and to help prevent inadvertent long-term storage.

Point 5: Visible Organization of Containers

Another important means of making the flow of delivered items smoother is to make the containers used for such items as clearly distinguishable as possible. We call this "visible organization of containers," which is part of the general idea of "visual control."

Figure 5.42 shows two examples of visibly organized containers, a parts tray and a parts box. These containers make it much easier for workers who select parts from them to understand which parts are which. They also make obvious which part has been overlooked, since the container should be empty when parts selection is finished. This also helps improve defect detection.

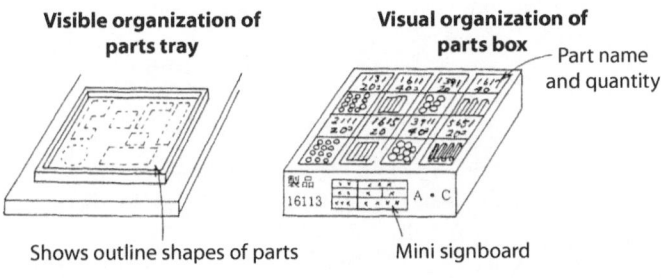

Figure 5.42 Visible Organization of Containers.

Chapter 6

Multi-Process Operations

Multi-Process Operations: A Wellspring for Humanity on the Job

Eliminating defects, raising the operating rate of workers and machines, and improving productivity are all matters of great importance in any factory. It is no exaggeration to say that higher productivity is the key to survival for companies today.

However, even "survival" is not reason enough to treat workers like machines. When you come right down to it, it is people—not machines—that make products. Productivity is important indeed, but not as important as respecting the humanity of our workers. Productivity and humanity must coexist in the factory. Sometimes, the two have conflicting purposes. If we raise productivity at the expense of humanity, we are doing ourselves a disservice in the long run.

For example, let us suppose that the workers in our factory each have very specific and specialized job tasks. One person hammers in bolts all day while another glues on labels. They have been doing this for five or ten years. How much pleasure do you suppose these workers derive from their work, and what sense of achievement or satisfaction have they gained after all those years?

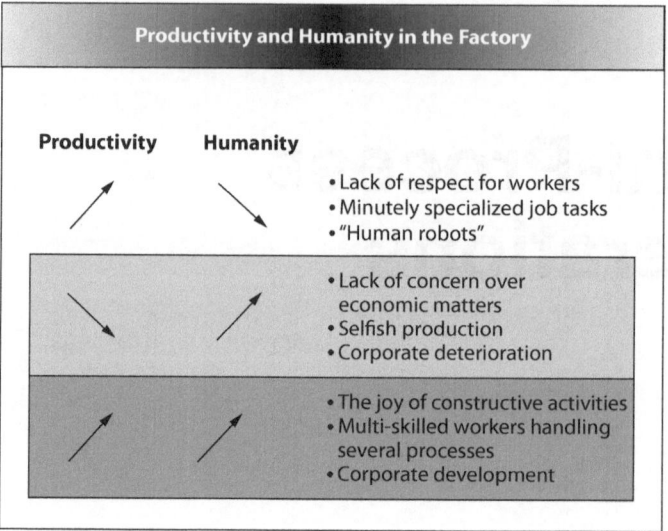

Figure 6.1   Relationship between Productivity and Humanity.

Now let us consider the opposite situation: a factory where humanity is respected even to the point where productivity is no longer important. When taken to such an extreme, humanity takes on shades of arrogance and, eventually, selfishness. Factories that take this path lose their vitality and ultimately fail.

Obviously, we need to find a way to satisfy both productivity and humanity (See Figure 6.1.)

The Difference between Horizontal Multi-Unit Operations and Vertical Multi-Process Operations

Building up one-piece flow production is the best way to get rid of defects, waste, and production delays. The basic concept in one-piece flow production is to send workpieces along the processing sequence one at a time, adding processing (value) to the workpieces at each process. As such, flow production is a very basic ingredient in JIT production. (For further description of flow production, see Chapter 5.)

The following are the main things we must have in order to establish flow production.

- *Equipment.* We need specialized machines that include only the essential required functions, are inexpensive, and are small enough to fit right in to the production line.
- *Equipment layout.* Equipment must be arranged according to the processing sequence. Workshops should be of the "flow shop" type (as opposed to the "job shop" type) and should preferably consist of U-shaped manufacturing cells.
- *Operational procedures.* We must give up "shish-kabob" production and learn one-piece flow in which workpieces are fed to and from processes one at a time. All workers must stand while working and learn to handle several processes in order to synchronize their work with the cycle time.
- *People.* We must train workers in the multiple skills they will need to handle several processes.

Multi-process operations are the key that opens the door to one-piece flow production. Without multi-process operations, there can be no JIT production system.

We are not likely to find much worker enthusiasm for multi-process operations if we introduce such operations in conventional "job shop" type workshops (workshops laid out according to function). Multi-process operations can be achieved in such workshops, but the amount of conveyance the workers would have to do themselves by walking and carrying workpieces makes it hard to find time for processing the workpieces. Therefore, we first need to change the equipment layout to the "flow shop" arrangement (equipment arranged according to the product). This changes the workshop from being a multi-unit process station to being a multi-process production line.

Obviously, we cannot change a multi-unit process station to a multi-process production line unless we change the equipment. A group of presses are only good for pressing and a group of drilling machines are only good for drilling. There is no way we can arrange multiple press units or drilling machine units into a multi-process production line. That is why we need to make the distinction between the grouping of machines that all serve a certain processing function (multi-unit process stations) and the grouping of machines that provide a sequence of processing functions needed to build a certain product (multi-process production line).

Figure 6.2 illustrates this distinction.

The concept behind multi-unit operations (that is, operations at multi-unit process stations) is to have one worker handle several processing machines that perform the same type of process. By contrast, the concept behind multi-process operations is to have one worker handle several processes (arranged according to the processing sequence).

No matter how many machines multi-unit operators handle, they only need one skill to operate them since the machines are all similar (presses, drilling machines, or whatever). Since multi-unit operations all take place at the same processing stage in the overall production line, we refer to multi-unit operations as "horizontal operations."

Conversely, operators who handle multi-process operations must have skills in several types of processes, such as presses, drilling machines, bending machines, and so on. We therefore refer to such workers as "multi-skilled workers." Since multi-process operations occur along a sequence of processes that include several stages along the overall production line, we refer to multi-process operations as "vertical operations."

Once we have established flow production that uses multi-process operations, we can be sure to expect higher quality. Almost all surface defects on products—such as dents, cracks, or missing parts—will disappear. One-piece flow will ensure that when the occasional defect does occur, the line

Horizontal operations

Process \ Product	A	B	C	D
1	◯	◯	◯	◯
2	◯	◯	◯	◯
3	◯	◯	◯	◯
4	◯	◯	◯	◯
5	◯	◯	◯	◯

	Multi-unit operations (horizontal)	Multi-process operations (vertical)
	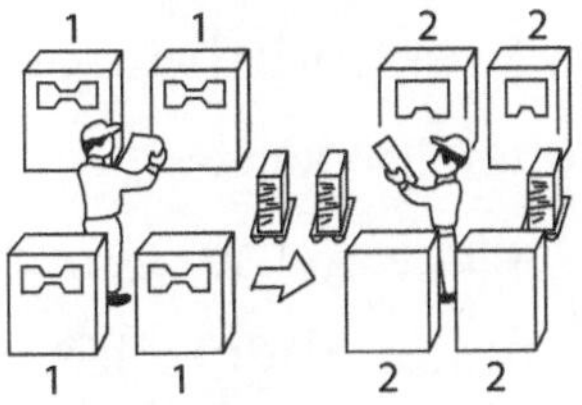 One worker handles four similar machines.	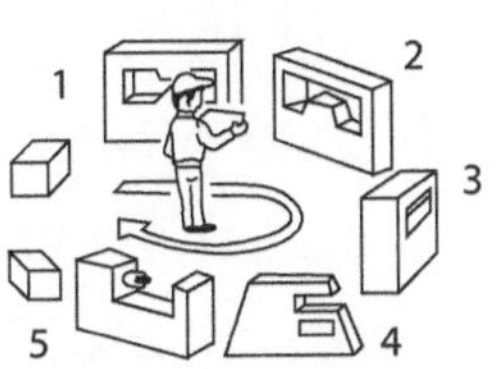 One worker handles five different processes.
Quality (Q)	• Dented, damaged, or missing items • Defective lots • Causes of defects remain a mystery • "I make the products, you inspect them." • Inspectors are responsible for sorting out all the defective products.	• Zero dented, damaged, or missing items • Zero defective lots • Causes of defects are tracked down and arrested. • Production workers do their own inspecting. • Quality is built in at each process.
Cost (C)	• Creates lots of waste related to in-process inventory, space, manpower, and conveyance • Costs vary depending upon volume. • Workshops try to save labor.	• Zero waste • Costs are steady regardless of volume fluctuation. • Workshops try to reduce manpower.
Delivery (D)	• Long lead-times • Chronically late deliveries • Not very adaptive to schedule revisions	• Short lead-times • Zero late deliveries • Adaptive to schedule revisions

Figure 6.2   Difference between Multi-Unit Process Station and Multi-Process Production Line.

can be stopped before an entire lot of defective products is turned out.

Best of all is the fact that this improvement enables us to track down the causes of defects and take appropriate countermeasures. In conventional shish-kabob production, anywhere from 500 to 1,000 defective units are produced before anyone notices the defect. Since the people who discover the defects are usually several stages down the line from the operators at the defect-causing process, it is very difficult to trace where that process is, and therefore it is very likely the defect will occur again.

By contrast, flow production using multi-process operations usually includes self-inspection by the multi-process operators. These operators not only turn out products, they objectively inspect them for defects. The inspection results reflect directly on their work and remind the operators that quality is built into products at each process.

In conventional shish-kabob factories, the general attitude among line workers is: "I just make them. It's up to the inspectors to inspect them." When we stop to think of the way the quality "buck" gets passed to the inspectors, we can recognize just how flawed the conventional approach is. The inspectors do what they can to sort out defects, but they do little or nothing to stop them from recurring.

We have been comparing shish-kabob production and flow production using multi-process operations only in terms of their quality aspects. But there are other important aspects, such as costs and punctual delivery. The cost impact of these two very different approaches includes the cost of in-process inventory waste, space-related waste, conveyance waste, and waste caused by putting things down and picking them up again. Flow production using multi-process operations can completely eliminate all of these kinds of waste.

One way to eliminate these kinds of waste is the practice of manpower reduction. (Chapter 7 describes manpower reduction in detail). Manpower reduction means using the minimum number of workers needed to produce the amount of products ordered by the client. When work is divided into single-skill tasks, more workers are needed to operate a production line and it is more difficult to reduce the number of workers when client orders shrink. Multi-process operations enable us to easily determine the minimum number of workers needed for any particular amount of output.

As for the delivery aspect, the lead-time for multi-process operations is remarkably shorter than for conventional shish-kabob operations. The former method not only prevents delivery delays, but reduces lead-time to where it is much better able to adapt to schedule revisions than the latter conventional method.

Questions and Key Points about Multi-Process Operations

Questions from Western Workers

Whenever I begin explaining JIT production to Europeans, Americans, and other Westerners, they usually look at me with a baffled expression, since their way of making things is so different from the way I am describing. After I describe multi-process operations to them, they pose questions that invariably include the following.

Question 1: Don't Multi-Process Operations Present Problems with the Labor Unions?

Yes. As a matter of fact, we can expect to have problems with the labor unions whenever we attempt to introduce multi-process operations in Western countries. In Japan, labor unions are "enterprise unions" in that each company has its own union. This means that companies can changeover to multi-process operations without having to change the union organization.

In the West, most unions are "craft unions." There are press workers' unions and lathe workers' unions and so on. The press workers' unions include people who specialize in operating presses, and this specialization makes it difficult, if not impossible, to introduce multi-process operations.

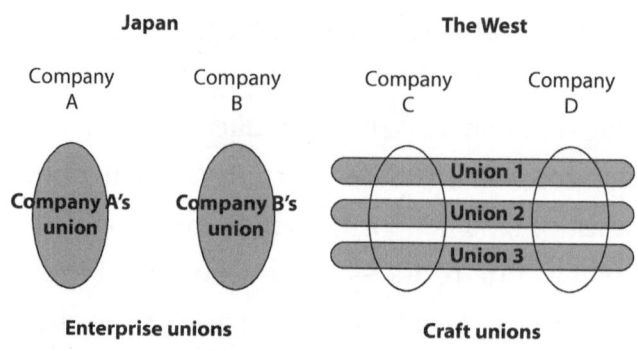

Figure 6.3   Enterprise Unions and Craft Unions.

When Japanese automakers build plants in the West, they generally try to hire all nonunion labor in order to facilitate the introduction of multi-process operations. (See Figure 6.3.)

Question 2: Do Workers Get a Raise in Pay after They Have Learned to Handle Multi-Process Operation?

There is a strong belief among Western workers that a worker's pay should correspond to the level of his or her skills. It would follow that someone who takes the trouble of learning the multiple skills needed for handling multi-process operations should expect a pay raise. In Japan, raises are generally tied to seniority in the company and not so much to specific skills. Very few Japanese workers or managers think that learning to handle multi-process operations should directly affect pay scales.

Question 3: If All Company Workers Need to Learn to Handle Multi-Process Operations, Wouldn't That Incur a Tremendous Amount of Training Costs for the Company?

In the West, it takes about three months of basic training to teach an unskilled worker how to operate factory equipment. Training the entire factory workforce to handle multi-process operations would indeed mean colossal training costs. But there are other, less expensive ways to train workers. In Japan, companies provide very little in the way of basic training courses for equipment operators. Instead, starting workers are given unskilled jobs and are required to spend about one hour of overtime each day just watching the skilled workers do their work. Another way Japanese companies keep training costs down is by thoroughly standardizing equipment so that few machines require a lot of specialized knowledge for their operation.

When seen from the perspective of the Westerners who typically ask the previous questions, it becomes obvious that JIT production is a very Japanese type of production. In particular, multi-process operations makes superb use of the flexibility in job assignments that characterizes Japanese companies.

Eight Key Points about Multi-Process Operations

Let us take a closer look at multi-process operations and the answers given to those three questions by examining the following eight key points about multi-process operations.

Point 1: Establish U-Shaped Manufacturing Cells

The first thing to do in preparing for multi-process operations is to abandon the "job shop" type of layout, which is appropriate only for shish-kabob production, and set-up a "flow shop" arrangement where the equipment is laid out according to the processing sequence. In other words, the various machines are lined up in a closely linked processing cell.

In this kind of cell, U-shaped lines are better than straight lines. Straight lines create waste by making operators walk farther when going back to get another workpiece at the end of each set of processes.

Figure 6.4 shows an automotive electronic parts assembly line. Before improvement, this line included about four or five cases of 24 parts each as in-process inventory between each set of processes. After improvement, they built a U-shaped

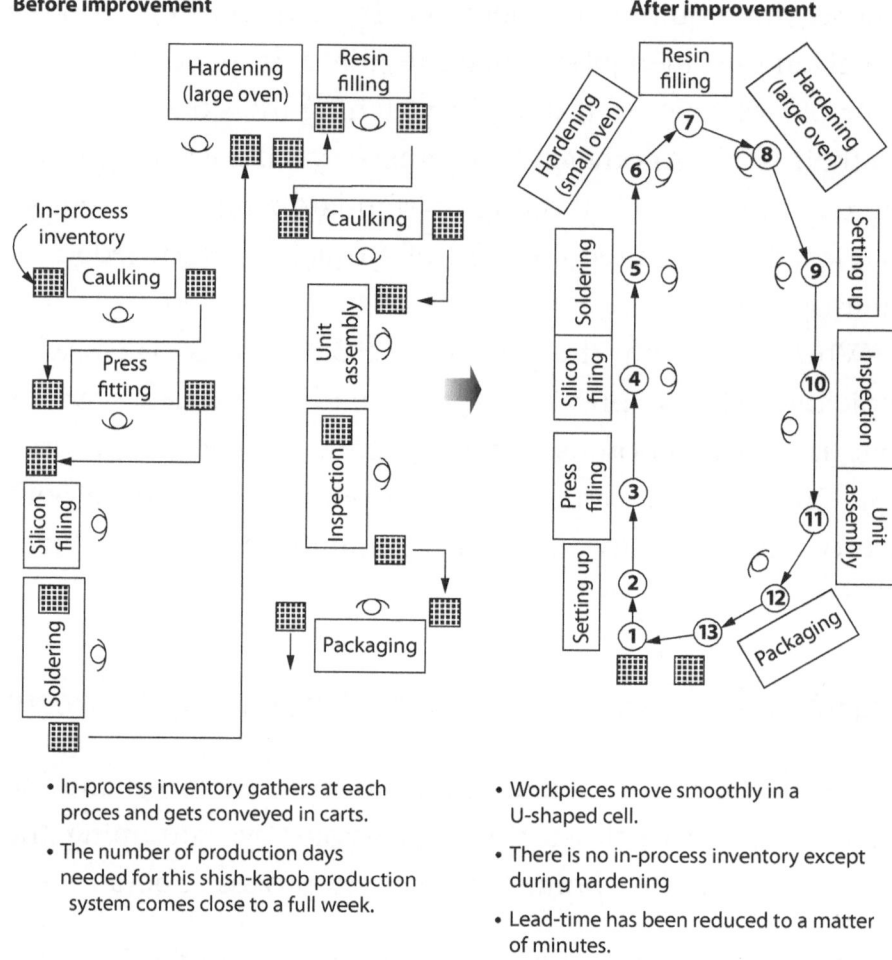

Before improvement

- In-process inventory gathers at each proces and gets conveyed in carts.
- The number of production days needed for this shish-kabob production system comes close to a full week.

After improvement

- Workpieces move smoothly in a U-shaped cell.
- There is no in-process inventory except during hardening
- Lead-time has been reduced to a matter of minutes.

Figure 6.4   Creation of a U-Shaped Manufacturing Cell for Automotive Electronic Parts Assembly.

manufacturing cell using a smaller hardening unit that could fit into the cell. This new layout eliminated cart conveyance and enabled a smooth one-piece flow of workpieces. The operators learned how to handle all 11 processes in the cell and, as a result, a smaller number of workers could produce the same output.

Point 2: Abolish Processing Islands

Manufacturing should have a steady rhythm to it, but who should determine the rhythm? The customers, of course. The rhythm that customer orders dictate is dictated first to the assembly stage, then to the processing stage, and finally

Before improvement

**Stem process
(Small processing "islands")**

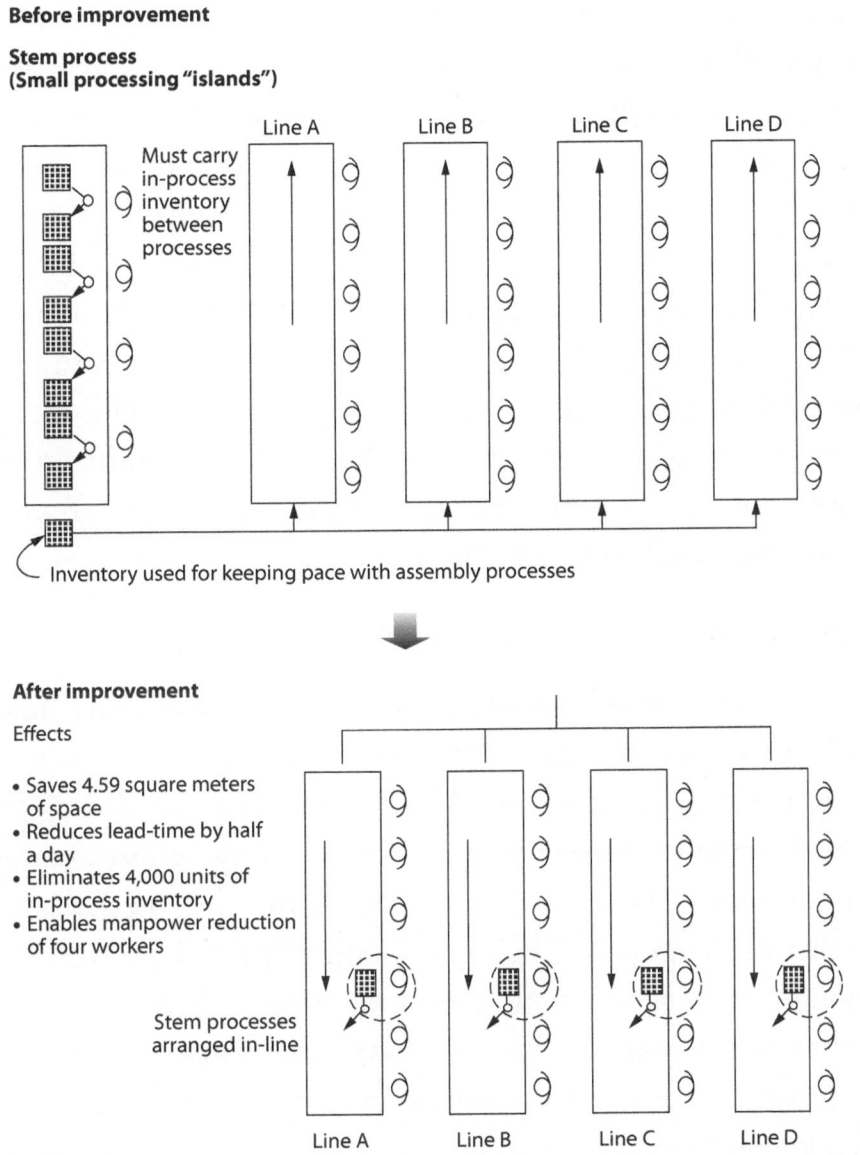

Figure 6.5   In-Line Arrangement of Watch Stem Processes (Eliminate Processing "Islands").

to the basic materials processing stage. However, at many factories, some processes exist independently as isolated little islands that run at their own rhythm. These little islands are full of waste—waste caused by their independent rhythms, by the resulting idle time for workers, and by their less obvious operating methods.

It is imperative to eliminate such processing islands and bring them directly into the line or cell. Figure 6.5 shows how a watch factory's winding stem process was brought into an integrated line. Before this improvement, the stem gear process was an isolated "island" that was operated at its own pitch by four workers, each of whom had to carry armfuls of inventory. They had to keep this little island well-stocked with workpieces in order to keep pace with the assembly lines.

After the improvement, they were able to balance this line with the assembly lines by including a stem process in each assembly line. As a result, they freed up 40.59 square meters of floor space, cut lead-time by half a day, eliminated the 4,000-unit stem inventory, and reduced the number of workers by four. (See Figure 6.5.)

Point 3: Make the Equipment Smaller

Usually, when a factory brings in new machinery, the major concern centers on how efficiently that machinery can be used. Even more important than the efficient use of any individual machines is the overall efficiency of the entire production system. (The concept of overall efficiency is discussed further in Chapter 2.) The equipment only needs to work fast enough to keep up with the cycle time. Therefore, we do not need fast, large, and expensive general-purpose equipment when the job can be done perfectly well using slower, smaller, and cheaper machines that perform only specialized tasks. Getting the right kind of equipment is the first step in bringing all equipment into a single line.

Figure 6.6 shows how a smaller shotblaster for automotive parts was developed. Before this improvement, this factory was using a shotblaster that was as tall as three people and was installed in its own room. Naturally, this machine lent itself to processing large lots, and the piles of in-process inventory in front of the shotblaster room took up twice as much space as the room itself. The shotblaster handled minimum lots of 500 units, and the units often banged into each other while being shotblasted, producing a defective rate of

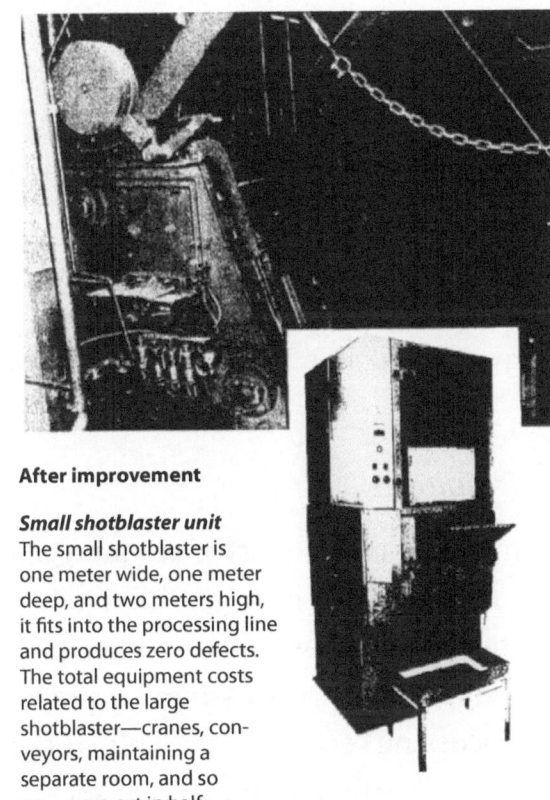

Before improvement

Large shotblaster unit
The large shotblaster was as tall as three people and had its own room. It processed lots of 500 units or more. Almost one-tenth of the shotblasted parts became defective due to collisions during shotblasting. Inspectors had to sort the defective units from the good ones after each shotblasting operation.

After improvement

Small shotblaster unit
The small shotblaster is one meter wide, one meter deep, and two meters high, it fits into the processing line and produces zero defects. The total equipment costs related to the large shotblaster—cranes, conveyors, maintaining a separate room, and so on—were cut in half.

Figure 6.6   Shotblaster for Automotive Parts.

nearly 10 percent. Inspectors were kept busy sorting out the defective parts after each lot was shotblasted.

The factory worked with the shotblaster manufacturer to develop a smaller machine that measured one meter wide, one meter deep, and two meters high. They called it the "one-piece shotblaster" Not only was this new shotblaster small enough to bring directly into the processing line, but it eliminated shotblast damage-related defects and removed the need for a shotblasting room, cranes, conveyors, space for in-process inventory, inspector manpower, and other forms of waste. (See Figure 6.6.)

Point 4: Standing While Working

At most home electronics or electronic component assembly plants, we can find rows of female workers seated alongside conveyors, busily assembling products.

Standing while working is a basic requirement for multi-process operations. Workers need to learn how to work on their feet. Once they are standing, they can more easily help their neighboring workers and thus eliminate idle time. Simply standing can do wonders.

Think about how the typical housewife fixes dinner. Can you imagine her seated at the kitchen counter or the stove, calling, "Dinner's almost ready," to her family as she busily prepares the food?

Point 5: Multiple Skills Training

Multiple skills training is an obvious necessity if we are going to have workers capable of handling multi-process operations. Multi-process operations occurs when a worker takes individual workpieces through the processing sequence, operating a variety of processing equipment. This differs from being an expert on any particular machine, such as thoroughly understanding the machine's design, retooling, operation, and maintenance.

The key to success in multi-process operations is simplifying the machines so that they perform only the essential processing function and do not require frequent fine-tuning. After that, we need to make certain that the workers learn how to systematically and confidently use the skills needed to operate those machines.

Figure 6.7 shows an example of multi-process operations at an auto parts machining line. This line is centered on numerically controlled machine tools and includes seven processes altogether. The operator is a 19-year-old woman. The key training points for multiple skills in this case included standardizing the machines, work procedures, and various other forms.

Point 6: Separate Human Work from Machine Work

This means making a clear distinction between work done by people and work done by machines, then separating the

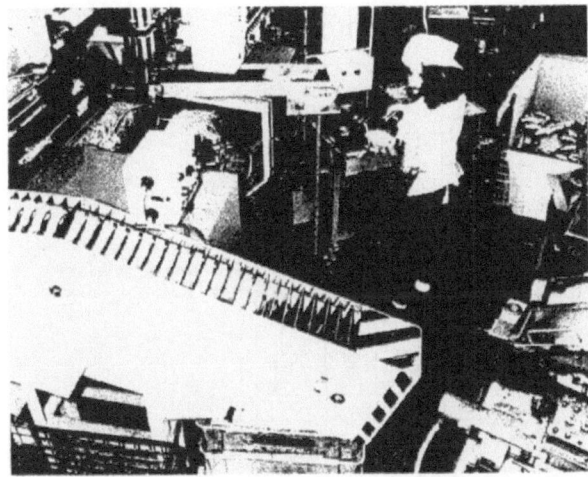

- This multi-skilled worker operates a machining line for automotive parts.

- This operater, a 19-year-old woman, has learned to handle seven different processes.

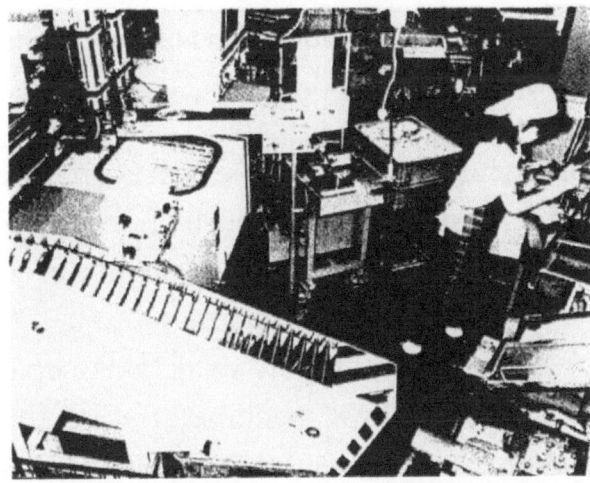

Figure 6.7   Multi-Skilled Worker in a Machining Line.

people from the machines whenever possible. (Separating human work from machine work is described further in Chapter 14.)

Usually, equipment operators stay close to their machines while the machines do their work. The fact is, however, that the worker and the machine each have separate tasks to do. Obviously, labor costs and equipment costs are both costs the company must pay.

If we can clearly distinguish between human work and machine work, the worker can leave the machine alone

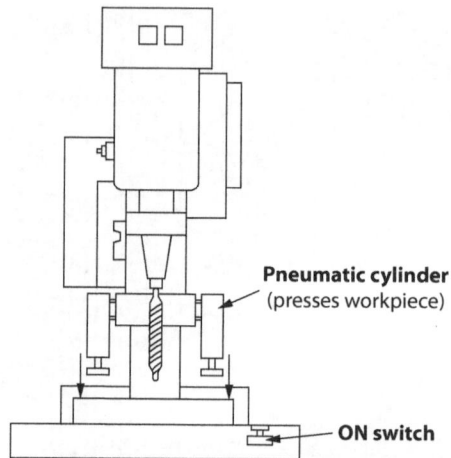

Pneumatic cylinder
(presses workpiece)

ON switch

- Before the pneumatic cylinders were installed, the operator had to hold the workpiece in position for drilling.
- After the improvement, the operator only needs to press the ON switch, and can then leave the machine alone.

Figure 6.8   Separating Human Work from Machine Work at a Drilling Machine.

to do its work while he or she goes on to the next human task. To make this possible, we must often develop devices and techniques that fall under the categories of "human automation" and *"poka-yoke."*

Figure 6.8 shows how human work was separated from machine work at a drilling machine. Before the improvement, the worker would press the ON switch and stand there holding the workpiece on the drilling machine with both hands. This meant that the worker was not free to do other work until the workpiece had been drilled.

After the improvement, pneumatic cylinders were installed on the right and left sides of the drill. When the worker presses the ON switch, these cylinders hold the workpiece in the correct position, enabling the worker to be completely separate from the machine.

Point 7: Human Automation and Poka-Yoke

Once the operator is able to let the machine do its own work, he or she is free to turn to the next human task. But what if

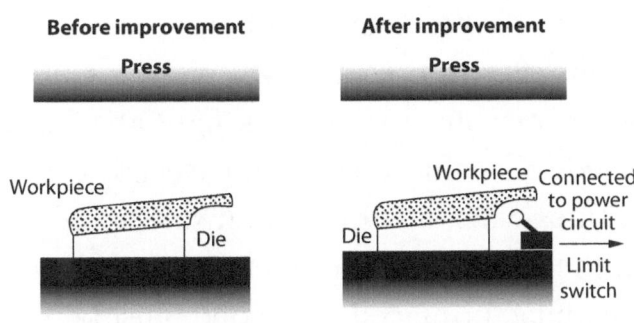

Figure 6.9   Use of a *Poka-Yoke* Device to Prevent Press Set-up Errors.

the machine starts producing defects without anyone there to notice? Does that mean we have to keep the operator there just to watch out for abnormalities? If we do that, we have not really separated the human work from the machine work.

Instead, we must come up with ways to prevent defects by having the machines automatically detect them and then stop operating. This is where human automation and *poka-yoke* come in. (Human automation and *poka-yoke* are described further in Chapter 14.)

Figure 6.9 shows a *poka-yoke* device that prevents set-up errors in a press. Before the improvement, the operator had to set-up the workpiece and then confirm correct set-up. Sometimes, however, the operator still made set-up errors, which resulted in defective products.

After the improvement, the machine was equipped with a limit switch that prevented the machine from operating unless the workpiece was set exactly right. This enables the operator to leave the machine without having to worry about the possibility of producing defective goods.

Point 8: Safety First

Once we have begun multi-process operations, we need to pay more attention than ever to safety matters. Everyone should remain mindful that "safety takes precedence over everything else."

One safety point is to keep start buttons separate from the machine themselves so operators will be at a safe distance at the instant the machines start working. Other useful safety devices include machine covers and electric eyes that shut off the machines when anything or anyone approaches them. Even when there are several operators working in the same U-shaped manufacturing cell, everyone must be very careful to maintain safety.

Precautions and Procedures for Developing Multi-Process Operations

Seven Precautions for Developing Multi-Process Operations

Single-skill workers are incapable of handling several types of processing machines and/or procedures. Therefore, we cannot have multi-process operations until we have taught the operators the wide range of skills they will need for the job. When training these operators, please note the following seven precautions.

1. *Make work procedures as simple as possible*
   There will inevitably be some cases where operators will need to learn certain procedures that take a long time to master or involve special skills. This is especially true of retooling and fine-tuning procedures.
   We can minimize these difficulties by simplifying work procedures so that anyone can easily understand how to perform them. In addition to simplification, thorough standardization can go a long way toward making multiple skills for multi-process operations easier to learn.
2. *Factory leaders should provide proper guidance*
   Effective leadership from factory managers and foremen is essential for ensuring swift progress in multiple-skill training.

After all, the operators are not the ones who best understand how the factory's various processes fit together and what procedures are involved in each process. Managers, foremen, and other supervisors have this knowledge and should put that knowledge to use in helping operators learn multiple skills for multi-process operations.

3. *Transparent operations*

When teaching work operations to a novice, we must explain the various operations and steps as clearly and fully as possible. This is what I mean by "transparent operations." (Chapter 13 explains the difference between transparent operations and standard operations.) To make our explanations transparent, we must uncover and elucidate all the little details that are usually considered "givens" and left unexplained. These "transparent operations" will ensure that even a completely unskilled worker will have all the information he or she needs to perform the job correctly.

Nothing should be left up to the factory's "oral tradition" of know-how that gets passed from person to person. Everything must be explicit and by the book. Job guidelines and operations manuals must contain clear descriptions of thoroughly standardized operations.

4. *Implement multi-process operations throughout the factory*

Multiple skills will soon deteriorate if they are only taught for certain processes or workshops. Company presidents and/or factory supervisors should put their full authority into promoting factory-wide multiple skills training. They should use whatever vehicles of communication are available to them (such as in-house newsletters and speeches) to issue progress reports on multiple skills training. They should also periodically hold "multiple skills contests" to present awards of recognition to the best trainees.

5. *Promote perseverance and set successive goals*

Multiple skills training needs perseverance like a car needs gasoline. Trainees have to be constantly encouraged to "hang in there" no matter what problems they encounter. There is no rush—the key is to take all the time you need to accomplish the training.

It is also very important to be systematic by clearly scheduling the various steps in multiple skills training. Draw up reference charts, such as a "Multiple Skills Training Schedule" or a "Multiple Skills Score Sheet," so that you can have an at-a-glance display of each trainee's progress.

6. *Make prompt equipment modifications*

Sometimes we need to modify equipment to make it easier for anyone to use or to enable the separation of human work from machine work.

Sometimes workshop employees get let down when production engineers or the equipment maintenance staff refuse to make the desired equipment improvements. It would be nice to have a team of equipment experts who specialize in JIT-related equipment improvements and are ready to work at a moment's notice. If the desired equipment improvement is simple enough, equipment operators or factory floor supervisors may be able to make the improvement themselves.

7. *Absolute safety*

Since multiple skills training requires novices to learn to operate various kinds of processing equipment, we must make sure the training is not hazardous. If even one accident or injury occurs during the training, it will likely have an adverse impact on morale and willingness to learn. We must therefore do everything we can to avoid all possible hazards.

Basically, two things can ensure absolute safety: careful safety checks during the design and operation of the equipment, and safety-minded discipline.

Five-Step Procedure for Training Multi-Process Workers

A few examples of multiple skills training can be found at just about any Japanese factory. Some factories proudly display banners or signs that announce their commitment to multiple-skills training.

However, almost all of these factories that promote multiple-skills training do not train workers to use these skills in a flow production system. Instead, they are mainly interested in having "pinch hitters" who can readily substitute for absent workers. These factories continue to operate shish-kabob production systems, and the multiple-skill workers are trained to move batches of workpieces from one process to the next in what I call "caravan style" operations.

They do not understand the true meaning of multiple-skill training and multi-process operations. Flow production forms the very foundation for JIT production. Factories must focus on the need to cultivate true multiple skills, which means the ones that are required for flow production using multi-process operations.

Multiple-skills training is a lot like small-group activities because it vitally depends on the involvement of the entire factory and on the encouragement provided by factory leaders. Many workers need to be prodded along—they are not fond of new adventures. They are snuggled safely into a cozy nest made up of work habits and the single set of skills they have practiced for years and years. They know their job perfectly and need not fear any unpleasant surprises. In fact, they can be confident and proud knowing that no one can perform their particular job as well as they can.

Multiple-skills training asks these seasoned "veterans" to throw away their single-skill achievements and start all over as amateurs. No wonder they resist so much.

We must use strong medicine to rid factories of this addiction to traditional work methods. We must go over the heads of section and division chiefs and include the company president and other top managers in the effort to encourage workers to accept the challenge of learning multiple skills.

I recommend following the steps described below when promoting multiple-skills training.

Step 1: Create multiple-skills training teams

It is usually best to follow the familiar format of small group activities by creating multiple-skills training teams. If the factory has already established a small-group activities program, it can simply set-up "multiple-skills training" as a new major theme within the program. The important thing is to help put trainees at ease and to set the stage for the challenge of developing multi-process operations.

Step 2: Clarify what the trainees' current skills are for each process

Before beginning the multiple-skills training, find out what skills and strengths the operator trainees already have and explicitly describe them. This can generally be done by entering the trainees' names on a chart and marking "skilled" or "unskilled" next to each process to indicate whether or not each trainee has the skills required for each process. You may need to make separate current ability marks when special skills are required in the process.

If possible, it would be even better to evaluate current skills using multiple levels instead of just the two levels of skilled and unskilled. A five-level skills evaluation might be organized as:

a. Level 1: Unable to do the operation.

b. Level 2: Able to do the operation if someone else does the set-up.

c, Level 3: Can generally do the operation, but needs minor guidance.

d. Level 4: Can do the operation well, except under unusual conditions.

e. Level 5: Can do the entire operation well.

Step 3: Use a "multiple skills training schedule"

We are now ready to set separate targets for each trainee whose current skills we just evaluated in Step 2. We should keep it simple by displaying person-specific lists of current conditions and targets, rather than process- or skill-specific lists. Also, we should avoid numerical indicators if more easily understood graphic ones can be used. Popular graphic display formats for this include "multiple skills score sheets" and "multiple skills maps." Figures 6.10 A, B, and C show three examples of multiple skills training schedules.

Step 4: Create a multiple skills training schedule that makes effective use of overtime hours and other opportunities

Once we have set specific targets for every worker, we need to set-up a multiple skills training schedule tailored to each worker's objectives. We should try to avoid using the noon hour, since that tends to disrupt production activities. It is better to use evening overtime hours.

Multiple Skills Training Schedule	○ Unable to do operation (LOSS) ◑ Can generally do operation (TIE) ● Can do operation well (WIN)	Factory name: HIC By: Yamasaki	Foreman: Yamasaki Date: 11/20/88

No.	Operator name	Printing	Mounting	Reflow	Cleaning	Visual inspection	Corrections	Soldering	Powder coating	Curing	Sealing	External view insp.	Tie bar card	Electrical char.	Packaging			Current date (11/30/88)	Target date (3/31/89)
1	Worker A	●	●	●	●	●	◑	◑	○	○	○	○	○	○	○	○	○	5 wins 7 losses 2 ties	12 wins 2 losses
2	Worker B	○	○		●	●	●	●	●	●	●	○	○	○	○	○	○	6 wins 7 losses 1 ties	12 wins 2 losses
3	Worker C	○	○	○	○	○	◑	◑	◑	●	●	●	●	○	○	○	○	4 wins 7 losses 3 ties	9 wins 5 losses
4	Worker D	○	○	○	○	○	◑	●	●	●	●	●	●	◑	○	○		7 wins 5 losses 2 ties	11 wins 3 losses
5	Worker E	○	○	○	○	○	○	○	○	○	○	●	●	●	○	○		3 wins 11 losses	7 wins 7 losses
6	Worker F	○	○	○	○	●	●	●	◑	○	○	○	○	●	○	○		3 wins 10 losses 1 ties	6 wins 8 losses
		○	○	○	○	○	○	○	○	○	○	○	○	○	○	○	○		
		○	○	○	○	○	○	○	○	○	○	○	○	○	○	○	○		

Figure 6.10A   Examples of Multiple Skills Training Schedule.

Multiple skills score sheet

Multiple Skills Score Sheet							○ **WIN**	
Period: 4/1/88 to 6/30/88			Gyochu Dept. 1, Section 2				△ **TIE** ✕ **LOSS**	
Operator name \ Process name	Pressing	Punching	Bending (1)	Bending (2)	Drilling	Finishing	**Wins and Losses**	
							4/1	6/30
Worker A	○	○	○	○	○	○	5 wins 1 loss	6 wins 0 losses
Worker B	○	○	△	✕	○	○	3 wins 2 losses 1 tie	4 wins 1 loss 1 tie
Worker C	○	○	△	✕	○	○	4 wins 2 losses	4 winss 2 losses 1 tie
Worker D	○	○	✕	✕	○	○	3 wins 3 losses	4 wins 2 losses
Worker E	○	✕	✕	✕	○	○	2 wins 4 losses	3 wins 3 losses
Worker F	✕	✕	✕	✕	○	○	1 win 5 losses	2 wins 4 losses

Multiple skills score sheet

Monthly check

							Section chief's check	
Multiple Skills Score Sheet							1　2　3　4　5　6	
Period: Dec.–Jan. 1988			Manufacturing Dept. 1, Section 2				7　8　9　10　11　12	
Operator name \ Process name	Coater 1	Coater 2	DB	PL	MJ	BP	CD	**Progress** 50%　　100%
Worker A	●	●	●	●	●	●	●	
Worker B	●	●	●	●	●	⊕	●	
Worker C	⊕	⊕	◐	●	●	◐	⊕	
Worker D	●	⊕	●	●	●	⊕	⊕	
Worker E	⊕	●	◑	⊕	⊕	⊕	⊕	
Worker F	●	⊕	⊕	⊕	⊕	⊕	⊕	

Evaluation criteria

⊕ Unable to do operation

◔ Able to do the operation if someone else does the set-up

◑ Can generally do operation, needs minor guidance

◐ Can do the operation well, except under unusual circumstances

● Can do entire operation well

Color coding

Black.................................1987 results
Red shading.....................Estimated 1988 results
Red.....................................1988 results

Figures 6.10B,C

For training in U-shaped manufacturing cells, it is best to pair up trainees with experienced workers and have them work together until they can keep pace with the cycle time. During this time, we will likely see the trainee and experienced worker develop a cooperative operations approach on their own.

Step 5: Periodically announce score sheet standings to raise worker awareness

At regular intervals, such as once or twice a month, factory supervisors should announce the trainees' current score sheet standings to make everyone aware of recent progress and to identify cases of delayed progress that need special attention. It is better to report the multiple-skills progress of trained teams rather than individual trainees.

If you choose to give progress reports for individuals, it is best to report their current status as "X percent of the way to the target," or in terms of "wins" and "losses" regarding specific skills (as shown in Figure 6.10).

And let us not forget the very important role the workshop leaders play in fostering multiple skills training. When learning a completely new skill, the trainee should begin by just watching an experienced operator or workshop leader do the job. These leaders in training have a direct and vital impact on the trainees. The trainees will learn the correct things, as well as any incorrect things, their more experienced colleagues demonstrate.

On-the-job training is clearly the best way to learn multiple skills for multi-process operations. Pulling a particular set of processes out of the production line to make an isolated island for training is not worth the time and trouble, since the training can be done within the production line.

In other words, training should be within the flow production system. This puts more pressure on performance. If we are just a little too slow, it causes problems for the next process. This keeps the trainees on their toes and aware of what is going on in the line. We call this method of training "multiple skills flow training."

Multiple skills flow training should take the following steps.

Step 1: Have the workshop leaders do the job first

Equipment operators learn quickly if given a chance to watch others do the job first. That is why it is best to start just by having them watch an experienced workshop leader do the job.

Step 2: Explain the operation points

Seeing is not enough. We also need to explain the procedures and main purposes of each job and make sure the trainees understand them thoroughly. At the very least, the teacher should explain the particular cycle time, operation sequence, standard operations, quality check points, and safety points.

Step 3: Hands-on practice

The trainee has seen and heard what he or she needs to know, it is time for some hands-on practice. The trainee should be allowed to attempt the entire set of operations for the process. If he or she starts lagging behind the cycle time, the trainer can step in to help. After repeated practice, the trainee will be able to perform the job according to the particular standard operations.

For example, let us suppose that a certain job includes five processes. The operator will start at the first process, and then in succession move on to the four others. If, at the third process, the trainee starts lagging behind the cycle time, the trainer should step in to help with processes 4 and 5. (See Figure 6.11.)

This works better than having the trainee just practice process 1 until he or she has learned it. The one-process-at-a time approach is too much like having isolated processing islands. The trainee will not gain a feel for flow production unless the training uses a flow production line of closely linked processes.

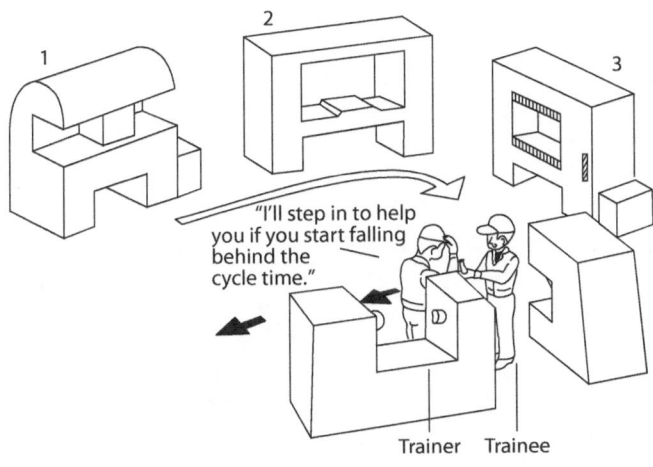

Figure 6.11   Trainer Helping Trainee during Multiple-Skills Training.

Step 4: Review the training immediately.

It is important to reserve a little time immediately after each training session to review the session. This is the perfect time to have another look at the key points in the operation and to resolve any confusion over what has been covered in that session. This should not be a purely negative review by the trainer. The trainer should always remember to praise the trainee. Beginners naturally feel uncomfortable around veteran operators, and the trainee's self-confidence is easily damaged. Harsh criticism is therefore often counterproductive. The trainer's responsibility is to bolster the trainee's confidence and enthusiasm. This is important.

In short, my advice to trainers of multiple skills for flow production is: show them, tell them, have them do it, then praise them. You need all four steps to get multiple-skill workers.

Chapter 7

Labor Cost Reduction

What Is Labor Cost Reduction?

The Approach to Labor Cost Reduction

Improvements in both productivity and humanity have long been major themes at factories everywhere. The market environment and needs differ from one era to the next, and factories must always attempt to make improvements in productivity and humanity that match the current market conditions.

Until recently, the general supply of products lagged behind demand, which in many cases meant, "If you can make it, it will sell." Factories sought to expand output volume, and looked at productivity-boosting measures as a means of doing just that. Human labor became more and more specialized, and factories tried to give workers simple tasks that they could master quickly. This simplification of worker roles as little cogs in a big machine tended to rob workers of the joy of creating things, but it served the factory's objective, which was to have a stable and highly regimented workforce that could turn out increasingly greater volumes of products. The following equation describes this volume-oriented approach to productivity.

$$\text{PRODUCTIVITY} \uparrow = \frac{\text{PRODUCTION OUTPUT} \uparrow}{\text{PRODUCTION INPUT} \rightarrow}$$

Eventually, the overall supply of goods overtook demand, leaving more room for diversification based on consumers' individual preferences. Manufacturers began to notice that their large production runs of identical products were no longer selling as briskly. Sales forecasts heralded the dawn of a new era, in which high volume output could no longer be assured of high volume sales.

Manufacturers began searching for a better way of making products that would sell. This was the advent of today's wide-variety, small-lot era. The soil was right for the JIT production system to take root. In contrast to the large-volume production approach that emphasized production and was thus a "product-oriented" or "product-out" approach, the new approach for the wide-variety, small-lot era emphasized the customers (that is, the market) and was a "market-oriented" or "market-in" approach.

Naturally, this new era saw growth in production volumes slow to a trickle. Manufacturers reckoned that the only feasible way to raise productivity in such a sluggish market climate was to reduce labor costs and other product input costs. They sought to cut labor costs by investing in greater mechanization and automation, but such improvements require a lot of investment funds and cannot ensure steady productivity because of rapidly changing market needs. Eventually, people started talking about building products more economically by matching production input to customer orders. This is the basic idea behind the labor cost reduction approach described in the following equation:

$$\text{PRODUCTIVITY} \uparrow = \frac{\text{PRODUCTION OUTPUT} \rightarrow}{\text{PRODUCTION INPUT} \downarrow}$$

Thus, we can define labor cost reduction as *meeting the needs (changes) of the next process (ultimately, the market) while incurring as few personnel costs as possible.*

Let us suppose, for instance, that a factory employs ten people to produce 1,000 units per month of product A. However, a recent slowdown in sales has shrunk customer orders to just 800 units a month. The traditional response to this situation is expressed in the following equations. The equation expressing the previous order level is:

$$\frac{\substack{\text{1000 UNITS} \\ \text{(Monthly output)}}}{\substack{\text{10 PERSONS} \\ \text{(Labor cost)}}} = 100 \text{ UNITS} \quad \substack{\text{(Number of products} \\ \text{produced per month} \\ \text{by each person)}}$$

The equation expressing the new order level is:

$$\frac{\text{800 UNITS (Monthly output)}}{\substack{\text{100 UNITS} \quad \text{(Number of products} \\ \text{produced per month} \\ \text{by each person)}}} = \substack{\text{8 PERSONS} \\ \text{(Labor cost)}}$$

The arithmetic is quite simple; assuming each worker can produce 100 units per month, the factory simply needs to reduce its workforce from 10 persons to 8 persons. However, it may not be so simple to reduce a ten-person workforce by two persons, especially if each of the ten workers specializes in handling just one type of machine.

This problem has forced some manufacturers to discard the concepts of single-process operations and strictly defined job roles and to instead embrace the new notions of multi-process operations and flexible job roles.

The realization of this kind of labor cost reduction is not without its technical obstacles, and the chief obstacle is a psychological one: giving up the fixed idea of large lot production.

The Difference between Labor Cost Reduction and Labor Reduction

Terms such as "labor reduction" and "labor savings" are familiar to all of us. We tend to think in these terms when confronted with the following types of situations.

Let us suppose that a factory has been using a single-spindle drill that required some manual assistance in drilling. Then the factory managers decide to buy a numerically controlled (NC) drill to automate more of the drilling work. However, the NC drill still requires a human operator, and so the factory is unable to reduce its manpower even after purchasing it. Whereas the worker used to be busy with manual drilling, now he or she simply sets up the workpiece, presses a start button, and watches the NC drill do the drilling. The NC drill has realized a labor savings (that is, the worker has less work to do), but not a labor cost reduction.

This case illustrates the meaning of the familiar term "labor savings." The investment in the NC drill has raised the plant investment cost without bringing a reduction in labor costs, so overall costs are actually higher than before.

Another familiar term is "staff reduction." Staff reduction means responding to demand fluctuations by simply reducing the number of workers without making any waste-eliminating improvements. However, if we just reduce the number of workers without making such improvements, the result will be labor intensification—in other words, more work to do for the remaining workers. Obviously, this kind of labor cost-cutting cannot go on for long. The following short definitions should help clarify the distinctions we need to make among labor reduction, staff reduction, and labor cost reduction.

- *Labor reduction:* Reducing the workload without cutting labor costs.
- *Staff reduction:* Reducing the workforce without removing waste (which means a heavier workload for remaining workers).
- *Labor cost reduction:* Removing waste, then using the minimum required workforce.

Labor Cost Reduction Steps

To be able to respond flexibly to changes in customer orders, we must have flexibility throughout our production system. Hence, the concept of "flexible production."

But exactly what needs to be made flexible? Everything—meaning every main element of production, from people and materials to machines, operating methods, and management. Let us look at these elements one by one.

- *People:* We can increase human flexibility by training single-skilled workers to become multi-skilled workers.
- *Materials:* We can improve flexibility in materials by moving from diverse specifications to shared specifications.
- *Machines:* Machines can in several ways be made more flexible by:
  1. Making nonmovable equipment movable.
  2. Switching from large machines to smaller ones.
  3. Switching from expensive machines to cheaper ones.
  4. Switching from costly "do-it-all" machines to cheaper specialized machines.
- *Operation methods:* Again, flexibility may be enhanced in several ways by:
  1. Abandoning lot production in favor of one-piece flow production.
  2. Switching from strictly defined job roles to flexible job roles.
  3. Switching from separate job responsibilities to cooperative job responsibilities.
  4. Giving up idiosyncratic operations and enforcing standard operations.
  5. Switching from "push production" to "pull production."
- *Management:* We can increase management flexibility by de-emphasizing statistical control and emphasizing visual control.

Thus, we need to make all sorts of changes to make the factory conducive to flexible production. Below, I have arranged some of these into a sequence of changes needed for realizing labor cost reduction.

Step 1: A Change in Philosophy

It is not possible to give up lot production and strictly defined job duties without also giving up the whole conventional "way of doing things." Even when someone decides, "OK, I'll give up all my preconceived notions about how things should be done," it is much easier said than done. Often, the old way of doing things is very old indeed; some workers have been doing things the same way for ten or even 20 years! The old way has become a deeply ingrained habit and cannot simply be cast aside. Workers who cannot bring themselves to admit the need for a change in philosophy might as well start preparing for retirement. Labor cost reduction requires flexibility, and flexibility must begin in the mind.

Step 2: Make Production Equipment Easy to Move Around

Large units of production equipment tend to have an imposing presence, as if they were standing with arms crossed and chest thrust forward, proclaiming, "I make widgets and I make them right here." We tend to lose our enthusiasm for making layout improvements when we come face to face with such huge machines that have usually been bolted to the floor. At such times, let us remember the following:

1. Whenever possible, install casters on equipment and work tables to make them movable. We must install the casters in a way that does not raise the height of these units.
2. If the machine has an oil pan under it, find out what is causing the oil leakage, fix it, then remove the oil pan and install casters.

3. Some machines have air ducts or power cords that limit their movability. In such cases, try lengthening the cord (make sure the length still meets safety specifications) and install flexible air ducts if possible.

Step 3: Get Rid of Processing Islands and Integrate Equipment into a Line

Labor cost reduction is not possible if workers are assigned to their own little isolated processing stations. We have to begin by bringing all those little islands together into one "land mass" so that workers can be grouped in one place. Once we have grouped our line workers, we can make a better line layout and start making improvements for one-piece flow production.

Step 4: Train for Multi-Process Operations Instead of Simple, Specialized Operations

The more we break production operations up into little pieces to be handled by different workers, the farther we get from labor cost reduction. Instead, we need to train workers in the multiple skills they need to handle multi-process operations. At each step of the way, we also need to implement thorough standardization.

Step 5: Standardize Equipment and Operations

Thorough standardization of equipment and operational procedures is essential for promoting multiple skills training. This training will progress much more rapidly if we can make the equipment easy enough for anyone to operate and the operations easy enough for anyone to perform.

Step 6: Level Out Production and Assign Appropriate Workloads

Find an average spread for product models versus volume, then divide this up by the cycle time and use the result as a basis for establishing standard operations. Use the cycle time to calculate the daily production output per person, then find the number of required workers depending upon how much each worker can do. (This procedure is described in detail in Chapter 10.)

When carrying out the above procedures, we must be careful to avoid putting too many workers on the line just because the workers are available. We must not ignore how much work each worker can comfortably handle. Workers are easily tempted to think, "Let's just take it easy since things are slow now." Managers tend to get lax about standards. Implement the 5S's and improvement activities to find out how much slack there is in the workforce and tighten up operations.

Points for Achieving Labor Cost Reduction

We must not make compromises when carrying out the above steps for achieving labor cost reduction. These steps include five salient points, which I list and describe below in the order of their appearance in the labor cost reduction steps.

- Develop flow production
- Cultivate multi-process workers
- Work in groups: no isolated workers
- Cooperative operations
- Separate people (from machines)

Develop Flow Production

Here are some typical characteristics of factories that are not conducive to flow production:

1. Equipment layout and operational methods are set-up according to the "job shop" model.
2. Equipment units are bolted in place and cannot be moved.

3. Each worker has distinct and strictly defined job duties.

4. People generally think large lots are better than small ones.

5. At processes where there is a lack of workers, workers are moved around "caravan style."

To begin changing from lot production (shish-kabob production) to one-piece flow production, we must do away with all of these obstructive characteristics.

Figure 7.1 shows an example of flow production on an assembly line for medical equipment. Before improvement, this line used eight workers, each of whom had a separate set of assigned tasks. This rigidity in task assignments made it nearly impossible to juggle operations when order levels fluctuated.

As part of the improvement, the layout was changed to accommodate flow production and operations were switched

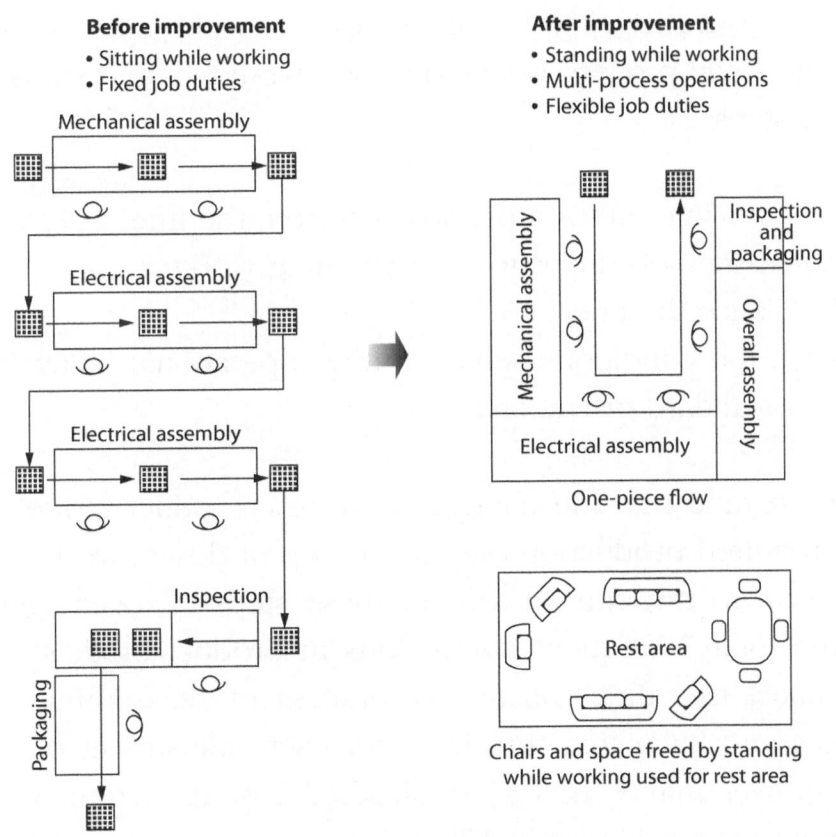

Figure 7. 1   Manpower Reduction through Flow Production (Medical Equipment Manufacturer).

from shish-kabob production to one-piece flow. The switch to multi-process operations not only enabled a labor cost reduction of two workers, but also made the line adaptable to ups and downs in order levels.

Before improvement, all of the workers sat while working. The improvement changed this to standing while working, which freed a lot of space. The extra space and unneeded chairs were used to make a rest area, which the assembly line had previously lacked.

Multi-Process Operations

To reduce the manpower required for a certain amount of production output, we first need to establish flexible job duties. Second, we must establish multi-process operations. This second step is the key to success in labor cost reduction.

If we were to try to reduce manpower without first establishing multi-process operations, we would have to follow these steps:

1. Removing one or more workers from the line.
2. Reassign job duties to the remaining workers.
3. Balance the line.
4. Set the conditions achieved after operational balancing as standard operations.

Each time the line changes to a new product model or the required production output goes up or down, we would have to go through all four of these steps all over again. Given today's frequent fluctuations in product models and volumes, this time-consuming process of reassigning job duties and balancing the line after each adjustment of the manpower makes this kind of labor cost reduction more trouble than it is worth. What factory managers are really wishing for is the kind of flexibility that enables them to easily reduce manpower one day to meet that day's output needs and to just as easily add manpower the next day. Only multi-process operations can make this wish come true, and that is why I call multi-process operations the key to successful labor cost reduction.

The three most important factors in establishing multi-process operations are:

1. Line workers must stop sitting and instead stand while working.
2. Lay out processes according to the processing sequence and make each worker take individual workpieces throughout the entire set of processes.
3. Set-up a company-wide multiple skills training program.

Once we have established one-piece flow using multi-process operations, the lead-time will be much shorter, and the shorter the lead-time, the lower the amount of in-process inventory.

Figure 7.2 shows how multi-process operations were established at a wood products factory's processing/assembly line.

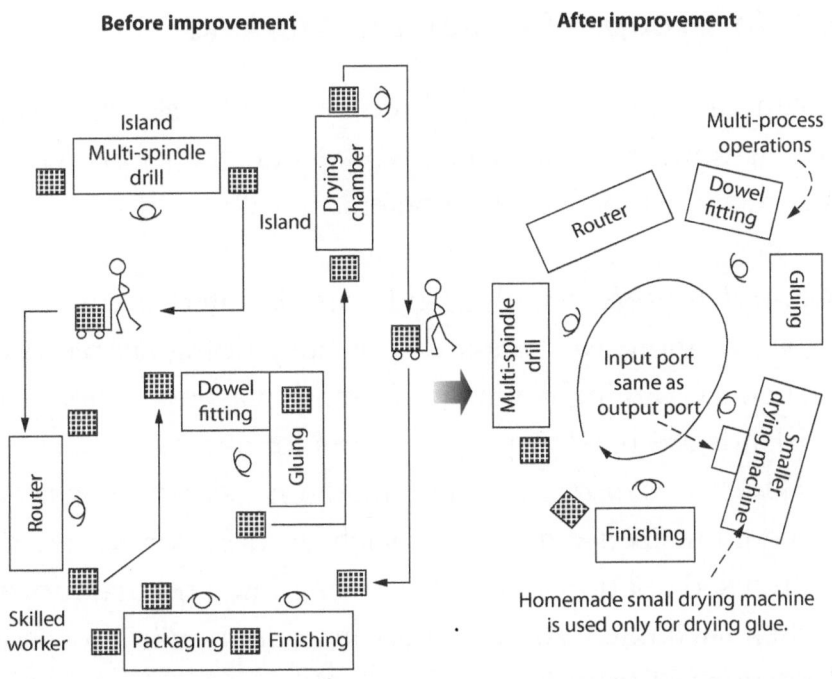

Figure 7.2 Multi-Process Operations for Processing Assembly Line at a Wood Products Factory.

Before this improvement, almost all work done by the line workers required special skills, and workers skilled in one process were rarely able to handle other processes. To change that, they first dramatically altered the layout, then trained their workers in multiple skills, and finally established one-piece flow using multi-process operations. They also made a very clever improvement in the drying process. Before, they had used a large drying chamber for drying glued parts. But since this chamber was too large for multi-process operations, they instead opted for a smaller machine that uses ordinary hand-held hair dryers and an auto-return device that returns the glued workpieces to the input site after they have been dried.

Not only did this improvement make the flow of goods on the line much more visible, it also made it easy to adjust the manpower to suit changing output requirements. It also helped get rid of waste, such as conveyance waste, caused by having isolated process stations.

Work in Groups: No Isolated Workers

We can distinguish among three types of "islands"—small medium, and large—at which workers do their jobs with no direct relationship to other workers.

- *Small islands:* Small islands are isolated areas where one or more workers are kept busy doing simple tasks, such as bagging items or mounting washers. Often, such islands are used to prepare parts for assembly.
- *Medium islands:* Usually, medium islands consist of medium-sized equipment, such as drills or lathes, that are used apart from the processing line and that move at their own pitch. As such, they are common in processing sections of factories.
- *Large islands:* Large islands generally include large equipment units, such as cleaning, coating, or welding

machines, all of which are designed for large-lot processing. Most common in processing sections, large islands are like dams that hold back the flow of goods. Sometimes, large islands require their own room or even their own factory facility.

If we have a large island, the most important point is to develop and make smaller equipment. If we have a medium island, we need to overhaul the layout and arrange the equipment according to the processing sequence. Finally, if we have a small island, our first step is to group the workers and assign cooperative tasks.

Figure 7.3 shows an improvement that was made at a household electronics assembly plant. Before the improvement, each worker worked separately at his or her own pace. Naturally, this imbalance resulted in a lot of waste caused mostly by operations, in-process inventory, and conveyance.

If we look at each worker involved in a small processing island, we can see the waste that is caused. But since the workers are separate, it seems there is nothing that can be done to improve the situation.

At the household electronics assembly plant, they began by setting up a conveyor and grouping all of the workers together. A conveyor can be valuable not only as a tool for maintaining a certain pitch, but also as a tool for grouping workers together.

After grouping their workers together, they laid out the various processes in order, then used the cycle time as a basis for assigning tasks. This helped eliminate the waste caused by having separate workers and also enabled a labor cost reduction of one worker.

Cooperative Operations

It is not at all unusual to have workers stand while working if they are working on processing tasks and using a lot of

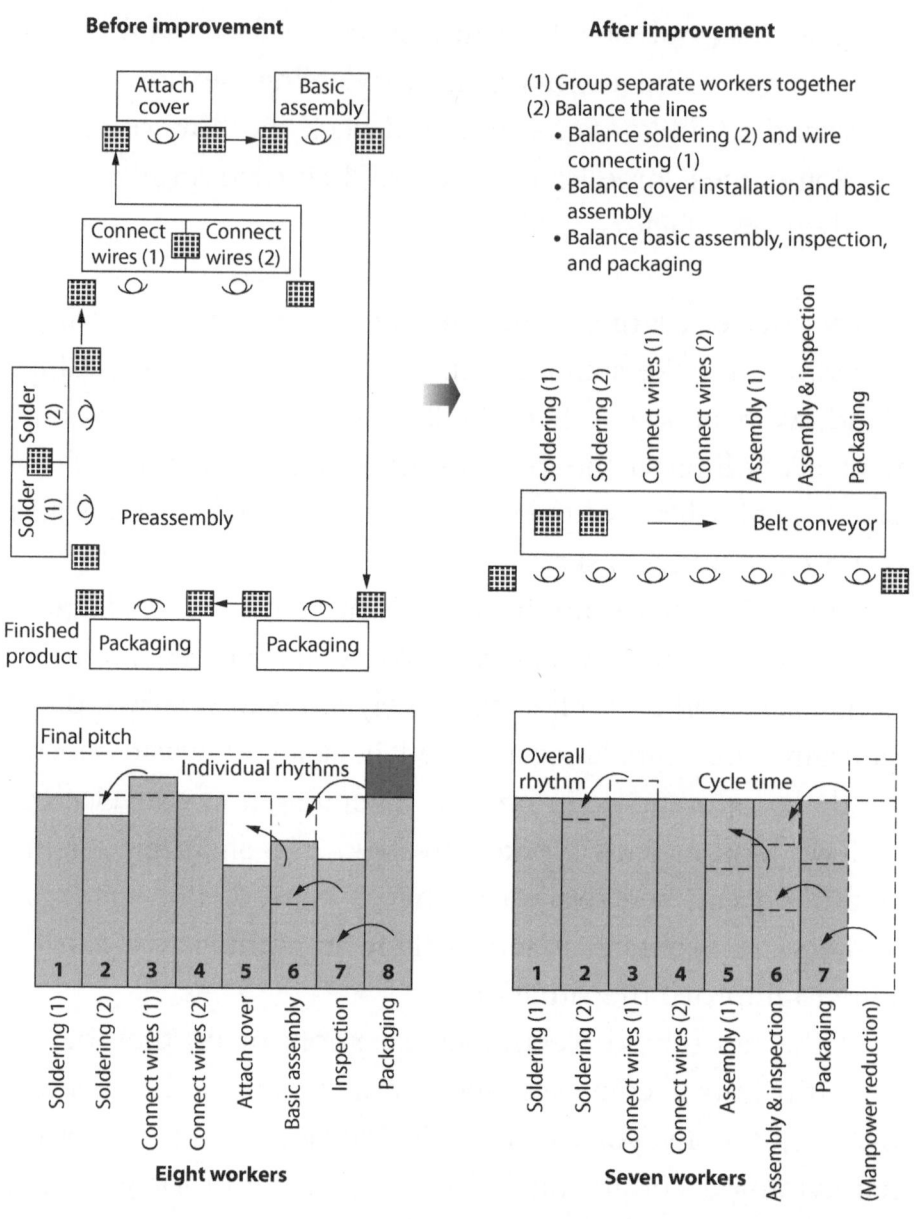

Figure 7.3   Manpower Reduction at Household Electronics Assembly Line.

machines, tools, and other equipment. In fact, it is hard to find seated workers doing this type of work. In assembly line work, however, the situation is almost the opposite.

Assembly line workers tend to plant themselves on their stools or benches and seem to believe they can get their jobs done perfectly without having to take one step. About the only time they use their legs is to join or leave the assembly

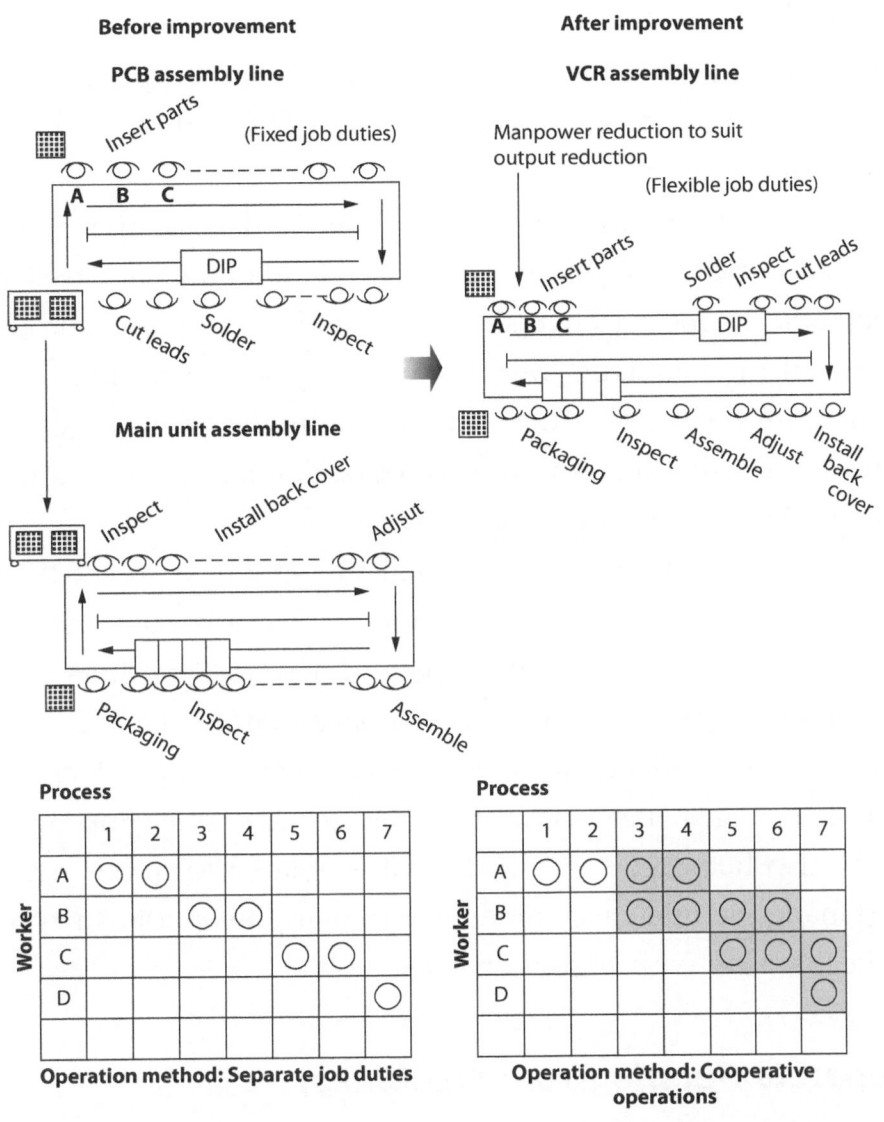

Figure 7.4 Cooperative Operations on a VCR Assembly Line.

line at the start or end of each shift and to get up for meals and breaks.

As long as work procedures are that rigidly established, labor cost reduction is impossible and it is even difficult to raise productivity. Before we can make any significant changes, we must establish the fundamental elements of cooperative operations, which are "standing while working" and "offensive (proactive) operations."

Figure 7.4 shows how cooperative operations and labor cost reduction were both realized at a VCR assembly line.

Before the improvement, the rigid task assignments made even slight increases in output something that required overtime work. Reductions in output were addressed by slowing down the pitch.

Because each line worker had his or her own strictly defined, separate tasks to perform, the line was not easily adaptable to model changes or fluctuations in daily output needs. If the managers were to remove just one worker (out of 61) in response to lower output requirements, they would have to take the time and trouble of balancing the remaining 60 workers on the line.

The answer, then, is to broaden the sphere of work that each line worker is responsible for, so that job duties overlap between neighboring workers and therefore workers can help their neighbor when he or she lags behind. This makes the line more adaptable to model changes and production output changes that occur from day to day. This improvement also helped get rid of the waste related to imbalances and made the line easily amenable to manpower adjustments in accordance with output changes.

Separate People (from Machines)

Most factory equipment operators are only rarely able to physically separate themselves from their machines and do other productive work while the machines are operating. The reasons for this unfortunate situation include:

1. Some of the processing activity requires assistance from the operators' hands or feet.
2. Operators have to set-up and retrieve workpieces manually from the machines.
3. Even when the operators do not have to touch the machines during their operations, they still must use their eyes and ears to detect defects or other problems.

4. Occasionally, operators are able to leave the machines completely alone, but only for a few seconds, so there is no significant separation.

5. Even when operators are able to leave the machines alone for significant lengths of time, there is nothing else at hand for them to do.

If the reason is any of the first three listed above, we need to develop some kind of device that will enable the operators to separate themselves completely—including their eyes and ears—from their machines. If the reason is the fourth or fifth one, we need to find them something more productive to do than just standing and watching the machines do their work. (Separating human work and machine work is described in detail in Chapter 14.)

Figure 7.5 shows how human work was separated from machine work in a printed circuit board (PCB) washing process. Before the improvement, the operator of this process had to insert the PCB manually into the washer and extract it manually after it was washed. Depending upon the timing

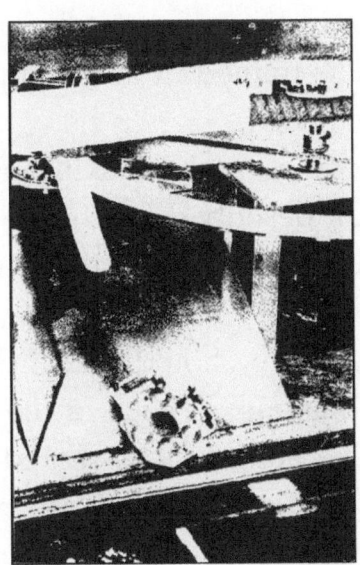

The operator used to insert and extract each PCB manually. The factory developed a human automation device that automatically extracts PCBs and sends them onto a conveyor. Now the operator only insert the PCBs.

Figure 7.5 Separation of Human Work and Machine Work at a Compact PCB Washer.

of the insertion, it could take several seconds until the PCB was ready to be extracted, during which time the operator was just standing by.

After the improvement, a human automation device was applied to the extraction step so that the operator no longer had to extract the PCB manually. Now, a shooter automatically moves the washed PCBs onto a conveyor line. After setting up each PCB in the washer, the operator can leave the machine alone and do other work.

Visible Labor Cost Reduction

Multiple Skills Training Schedule

Multi-process operations are the most decisive factor in achieving labor cost reduction. Once all workers have been trained for multi-process operations, it is a cinch to move workers around and to add or subtract workers to suit current manpower needs.

While this method known as multi-process operations is vital to such flexibility, it is the operators themselves who make it a reality. In other words, the key point for labor cost reduction is to have all workers trained in the multiple skills needed for multi-process operations.

Multiple skills training schedules, multiple skills maps, and multiple skills score sheets (all described in Chapter 6) promote progress in multiple skills training by making the training more visible.

The following are five steps we should take in training workers for multi-process operations. At each of these steps, we need to reaffirm a positive attitude that should include the three "P's": Painstaking care, Patience, and Perseverance.

■ Remember the three "P's": Painstaking care, Patience, and Perseverance.

■ The five steps in multiple skills training are:
1. Find a way to describe and/or illustrate the workers' current skill levels so that anyone can understand them.
2. Once or twice a year, evaluate and display progress in multiple skills training.
3. Make up a schedule of skills achievement targets.
4. At weekly, biweekly, or monthly intervals, mark the results that indicate progress toward achieving skill targets, and announce these results at meetings or other appropriate occasions.
5. Some trainees may find certain processes difficult to master. This is when the workshop leaders need to step in and provide moral support and extra training.

Labor Cost Reduction Display Board

In assembly lines, the first parameter to keep track of is the pitch time (otherwise known as the cycle time). We must at least keep track of the line's rhythm: How many units are we turning out per day and does this match the current volume of orders? This information is so vital that it should always be available to us at a glance.

If we want to improve the range of immediately available information, we should also include an up-to-date display of labor cost reduction parameters. In other words, how many workers does the line currently require? It is helpful to have that information around to check at any time.

Figure 7.6 shows a "labor cost reduction display board" that can serve just this purpose.

Once we know how many units each worker can reliably turn out in a day, we divide the day's total output by that number of units to obtain the minimum number of workers needed for the day. For instance, let us suppose that each worker on the assembly line can assemble 100 units a day:

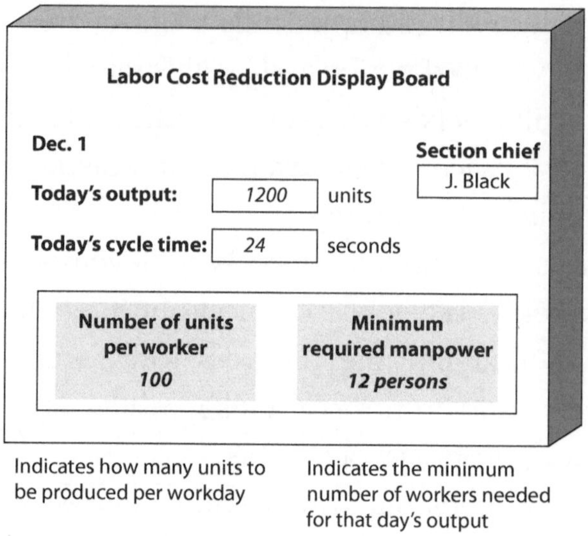

Labor Cost Reduction Display Board

Dec. 1

Section chief

J. Black

Today's output: 1200 units

Today's cycle time: 24 seconds

Number of units per worker	Minimum required manpower
100	12 persons

Indicates how many units to be produced per workday

Indicates the minimum number of workers needed for that day's output

Figure 7.6 Labor Cost Reduction Display Board.

1. If the total output is 1,000 units, the number of required workers is 10, and the pitch time is 28.8 seconds per unit (based on an eight-hour workday).
2. If the total output is 1,200 units, the number of required workers is 12, and the pitch time is 24 seconds per unit (based on an eight-hour workday).

It is good to keep a labor cost reduction display board (such as the one shown in Figure 7.6) posted in a conspicuous place so that everyone at the assembly line can quickly refer to it at any time.

Chapter 8

Kanban

Differences between the *Kanban* System and Conventional Systems

The Reordering Point Method and the Kanban *System*

Many people think the *kanban* system comprises the central technique around which JIT production is built. Let it be understood, however, that *kanban* are just one of several tools used to maintain JIT production and are by no means a central aspect of the JIT production system.

It has been said, "Wherever there are *kanban,* there is in-process inventory." *Kanban* and in-process inventory are indeed very closely related to each other. We can find *kanban* circulating here and there all over many Japanese factories. Because the *kanban* are in such conspicuous use, the factory workers imagine they have established JIT production in their factory. From the perspective of true JIT production, one might ask, "Why use *kanban?*" There is no reason why *kanban* should be absolutely necessary for every JIT production system. Rather, the essential thing in JIT production is a healthy flow of goods. The *kanban* system is not even an original idea, really. It is something that grew out of a statistical inventory management method known as the reordering point method.

As its name suggests, the reordering point method enables factories to reorder the same volume of parts or products each time. When the inventory amount drops to a certain level (the reorder point), another order is issued for the same amount as before to replace the depleted inventory.

Let us examine a list of the reordering point method's chief characteristics:

■ It enables inventory to be managed without having to pay attention to demand fluctuations.
■ It is not suitable when sharp demand fluctuations are typical.
■ It helps keep inventory management costs down.
■ It is conducive for use in an automated reordering system.
■ It helps lighten the clerical workload.

In view of the above characteristics, we can conclude that the reordering point method is a good inventory management method when the inventory consists of products having the following three characteristics:

1. A stable consumption volume
2. Easy to purchase and easy to store
3. Relatively inexpensive

We should regard the reordering point method's unsuitability for products whose market demand fluctuates sharply as the method's most important characteristic. This means, of course, that this method is only suitable for managing inventory of products that have stable demand.

We should also note that the exact same problem exists for the *kanban* system: If demand has large and unpredictable ups and downs, even the *kanban* system will not prevent product shortages or gluts. At the production planning stage, we can spread out the various product models and volumes

		Reordering Point Method	***Kanban* System**
Similarities		1. Enables inventory to be managed without paying attention to demand fluctuations 2. Not suitable when sharp demand fluctuations are typical 3. Helps keep inventory management costs down 4. Conducive to use in an automated reordering system	
Differences	**Information and goods**	Information and goods are kept separate from each other (inventory [= goods] is managed according to the warehouse entry/exit vouchers [= information]).	Information (*kanban*) and goods are kept together.
	Management	Requires constant inventory management (warehouse entry/exit management)	Does not require management
	Visual control	Does not enable visual control	Enables visual control
	Relationship with factory	Managed separately from the factory	Closely related to the factory and factory operations
	Relationship to improvement activities	None	Decreasing numbers of *kanban* indicate a need for improvement.

Figure 8.1   Similarities and Differences between the Reordering Point Method and the *Kanban* System.

and average them out. This is called "level production." (Level production is described further in Chapter 10 of this manual.)

If we use level production to help minimize waste, we are no longer able to manufacture products in large batches or lots. Therefore, factories that rely mainly on lot production or batch production need to use a rather strict production method. Figure 8.1 lists some of the similarities and differences between the reordering point method and the *kanban* system.

Conventional Production Work Orders and the Kanban *System*

Conventional production work orders indicate the type of production to be carried out at each process based on process-specific operation plans that have been developed as part of the overall production schedule. This means that each process relates vertically to the production schedule and not

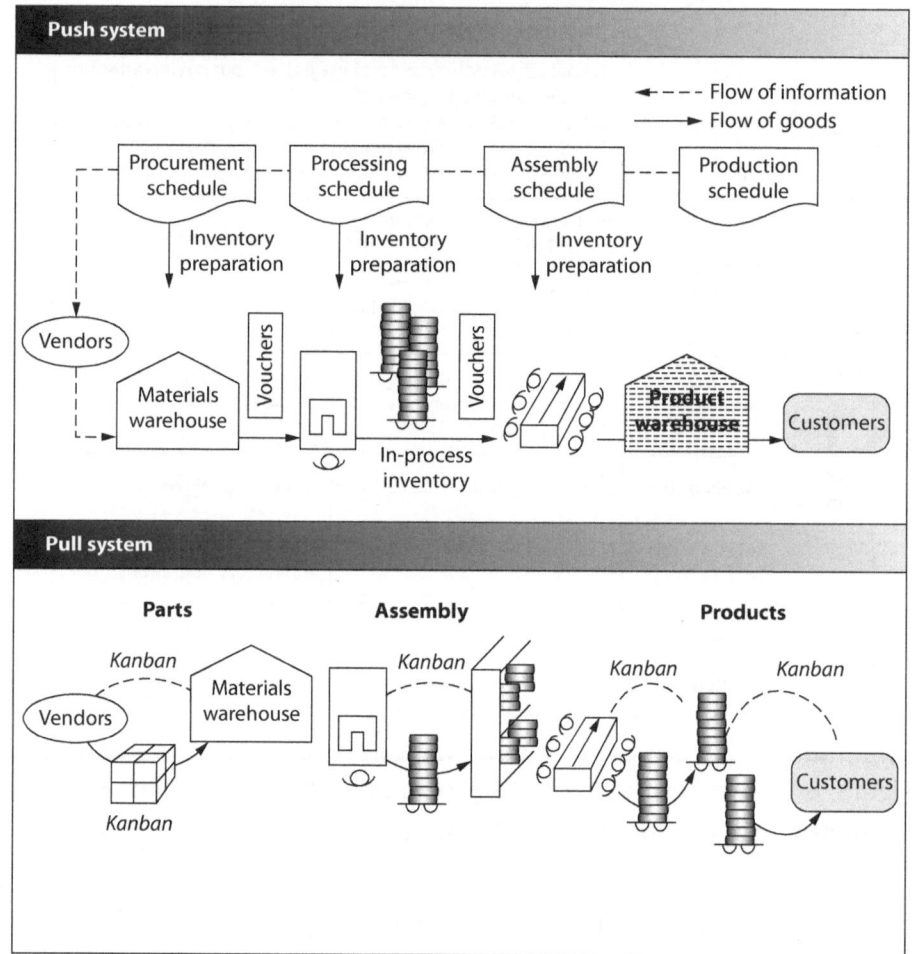

Figure 8.2 Differences between Conventional Work Orders and Kanban.

horizontally to other processes. Nevertheless, production is still a series of processes.

Usually, factories that use conventional production work orders also use the "push system" in which the upstream processes take priority over downstream ones in terms of how goods are moved and controlled between processes. By contrast, the "pull system" is a basic principle of the *kanban* system. As shown in Figure 8.2, the pull system means that downstream processes fetch from upstream processes only the goods that are needed, only when they are needed, and only in the required amounts. Naturally, as an upstream process is depleted of its products, it "pulls" more workpieces from the previous process, and this gets repeated as a chain reaction all the way up the line.

Figure 8.2 illustrates some of the differences between the push system and the pull system. The push system emphasizes the flow of information in that it "pushes" or "imposes" the production schedule and the in-process inventory onto downstream processes. In the pull system, *kanban* are attached to in-process inventory, so that when goods are pulled from a process by the next process, the item indication on the *kanban* can serve as a work order for the previous process.

The biggest difference between the push system and the pull system is the way information relates to actual goods. While the push system deals primarily with general production-related information first, after which production flow occurs as a result, the pull system deals with process-specific information and the transfer of goods. The pull system therefore makes it easy for changing conditions in downstream processes to impact upon upstream processes. The push system tries to stubbornly fulfill the original production schedule no matter what is going on downstream. This rigidity is reflected in the unchangeable nature of the typically "confirmed" production schedule for the next week and the "estimated" production schedule for the following three weeks. Even if the flow of goods in the factory should change drastically from what was envisioned when the production schedule was created, the inventory brought in for that schedule is still imposed upon downstream processes regardless of its actual value under the changed situation.

By contrast, the pull system dictates that as soon as clients order certain products, work orders for those products are sent to the assembly line, which in turn orders the parts it needs for those products from the processing line. The processing line then orders from the materials procurement people, and so on. This means that order information (that is, *kanban*) travels upstream from sales to assembly, instead of downstream from planning to materials procurement. This makes for a very flexible production system.

Functions and Rules of *Kanban*

Functions

As I said earlier, *kanban* comprise a tool for establishing and maintaining Just-In-Time production. As such, it is similar to the autonomic nervous system. When some kind of problem occurs at a downstream process, the system has a function for alerting upstream processes and stopping the production line.

In other words, *kanban* have two main functions.

Function 1: To Act as an Autonomic Nervous System for Just-In-Time Production

Kanban pass along information about downstream conditions to upstream processes, just as the autonomic nervous system notifies the brain of stimuli encountered by the body's peripheral nerves. This function can be broken down into two main roles.

1. *To provide pickup and work order information.* In this role, *kanban* provide two types of information: data about which items have been used and in what amounts, and also instructions on where and how certain items are to be manufactured.
2. *To eliminate overproduction waste.* In the *kanban* system, production occurs when goods are pulled from upstream processes. Otherwise, no production occurs. This is what makes the *kanban* system a "pull system."

Function 2: To Improve and Strengthen the Factory

As long as *kanban* are used as information, they remain attached to the goods that they give information about. As such, *kanban* serve beautifully as a visual control tool. This function of *kanban* also plays two roles:

1. *A tool for visual control.* Conventionally, production-related information is issued first, and the actual goods

come into play later on. In the *kanban* system, though, the information arises as a result of the consumption of goods. Therefore, *kanban* are always used with actual goods. And the way (including the order) in which *kanban* are eventually detached from goods shows us an obvious indication of how factory operations are proceeding and which goods in the flow of goods are receiving the highest production priority. This makes *kanban* an excellent tool for visual control.

2. *A tool for promoting improvement.* Inventory tends to conceal problems in the factory. Similarly, an overabundance of *kanban* indicates there is too much slack in the in-process inventory. Reducing the number of circulating *kanban* can help reveal the problems that can remain hidden under such slack conditions.

Rules

As mentioned above, *kanban* are the factory's autonomic nervous system and are a tool for building a stronger, healthier factory. The following six rules must be observed if we intend to make the most of *kanban*'s potential for factory improvement.

Rule 1: Downstream Processes Withdraw Items from Upstream Processes

Rule 2: Upstream Processes Produce Only What Was Withdrawn

Upstream processes must always produce in direct relation to downstream production. In other words, the previous process should produce only what was needed by the next process, only when needed, and only in the amount needed.

Rule 3: Send Only 100 Percent Defect-Free Products

Quality is built in at each process, and processes should never send any defective goods downstream. Passing the quality buck not only creates confusion at downstream processes, it also conceals problems at the defect-producing process and ultimately brings disorder to the entire factory.

Rule 4: Establish Level Production

Production leveling is a method that eliminates variation in flow at different processes and helps maintain stable, smooth production. (See Chapter 10 for a detailed description of production leveling.) This is different from the kind of balancing of load that occurs in a shish-kabob production system when using a planning method called Capacity Requirements Planning (CRP). Rather, it is the thorough balancing of product models and volumes within the framework of the production schedule.

Rule 5: Workshop Indicators

Kanban should also move with the goods to ensure visual control.

Rule 6: Use Kanban *to Discover* Needs for Improvement

By gradually decreasing the number of *kanban* in circulation, we can better reveal missing items and line-stopping problems, which we need to follow up with causal analyses and improvement measures.

How to Determine the Variety and Quantity of *Kanban*

Types of Kanban

First of all, let us be sure we understand the distinction between *kanban* and the signboards that describe where things are placed in the workshop. The latter are the manifestations of the "signboard strategy" that serve to make orderliness—one

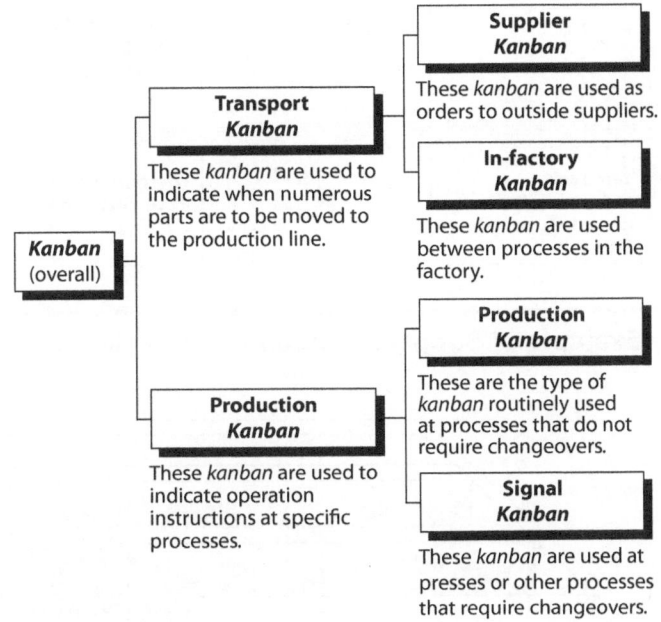

Figure 8.3 Types of *Kanban*.

of the 5S's—more visible. (See Chapter 4 of this manual for a description of the signboard strategy.)

Since the Japanese word *kanban* corresponds to "signboard" in English, *kanban* and signboards can be easily confused. In this manual, we use the English word "signboard" when discussing the signs used in the signboard strategy and the Japanese word "*kanban*" when discussing the signs attached to in-process inventory that comprise the factory's autonomic nervous system.

There are as many types of *kanban* as there are types of *kanban* applications. Figure 8.3 classifies these *kanban* types according to their functions.

Let us look at these *kanban* types in more detail.

Type 1: Supplier Kanban

Also known as "parts-ordering *kanban*," these *kanban* are used to order large numbers of parts that need to be delivered to assembly lines. Often, such *kanban* are sent to outside suppliers who deliver the parts on demand (see Figure 8.4).

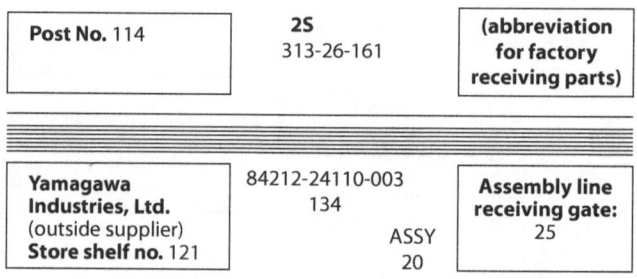

Post No. 114	**2S** 313-26-161	(abbreviation for factory receiving parts)
Yamagawa Industries, Ltd. (outside supplier) **Store shelf no.** 121	84212-24110-003 134 ASSY 20	**Assembly line receiving gate:** 25

Figure 8.4   Example of Supplier *Kanban* for Outside Supplier.

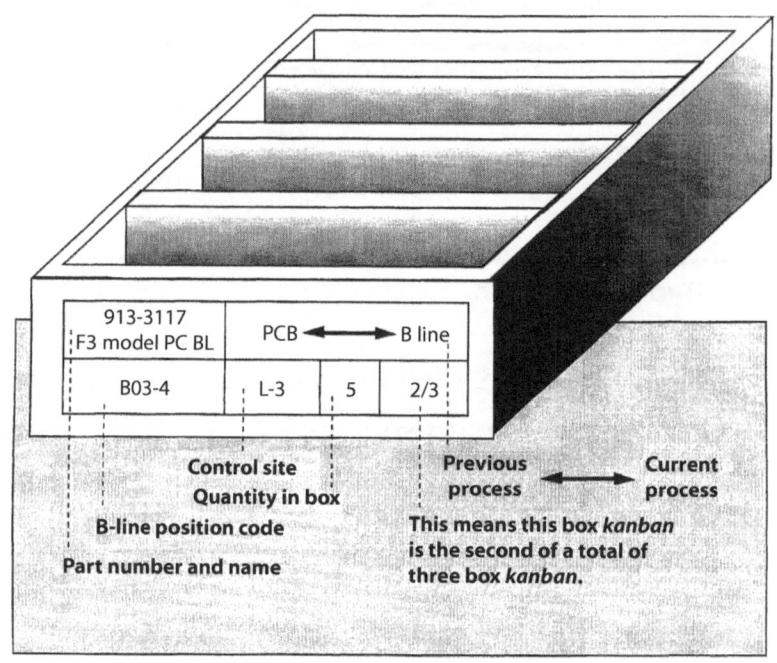

Figure 8.5   Example of In-Factory *Kanban*.

Type 2: In-Factory Kanban

Assembly lines also use parts that are processed and delivered from within the same factory. In-factory *kanban* are used to order such parts from upstream processes. Thus, they are also known as "pickup *kanban*" or "withdrawal *kanban*." (See Figure 8.5.)

Sometimes, in-factory *kanban* are used even when only one part is being withdrawn, or they can be used as "sequential withdrawal *kanban*" for when parts must be supplied in a certain order for assembly. The types of in-factory *kanban*

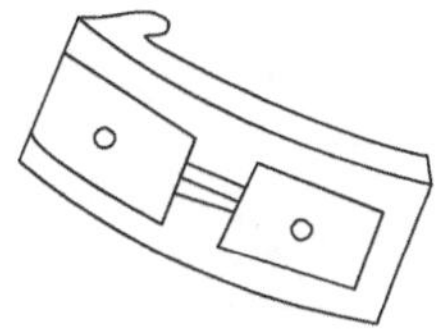

		Previous process ⟵⟶ Current process	
	Process	Plating (ME-47)	Coating (TO-13)
	Part name	51341-162600-00 Tail lamp rim	
	Capacity	20	
Control no. L-2	**No. issued**	6/10	

Figure 8.6 Example of Production *Kanban*.

can range from ordinary plates to "box *kanban*" (attached to boxes) and "cart *kanban*" (attached to carts).

Type 3: Production Kanban

Production *kanban* are used for in-process inventory within processes. These are the type of *kanban* most people think of first when *kanban* are mentioned in an overall sense. Usable in either specialized or nonspecialized lines, production *kanban* give instructions on operations at each process that does not require any (or hardly any) changeover time (see Figure 8.6).

Type 4: Signal Kanban

Moving some types of equipment (such as presses) directly into the production line can be difficult due to the costs involved. In addition, when model changes occur, the changeover procedures for such equipment can be quite time-consuming. As a result, lot production is sometimes unavoidable, at least at processes using these kinds of equipment. Signal *kanban* are used for such lot-production situations. (See Figure 8.7.)

How Many Kanban Do You Need?

Kanban help maintain level production. They also help maintain stable and efficient operations in which the same

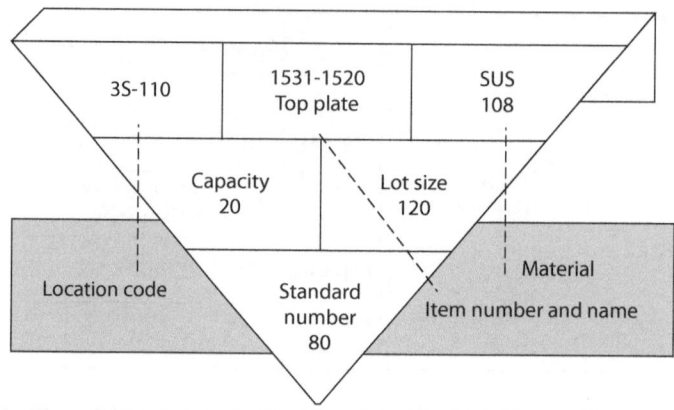

Figure 8.7   Example of Signal *Kanban.*

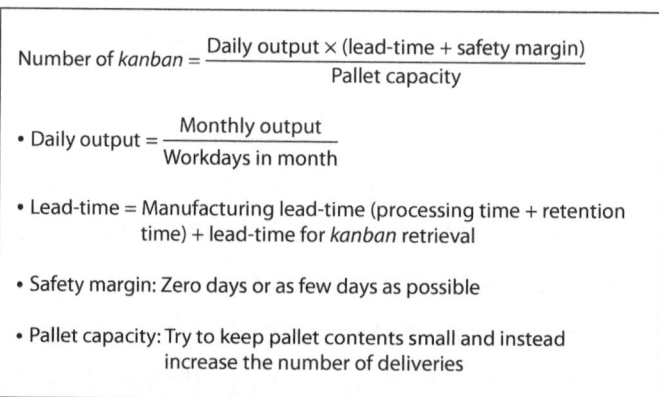

Figure 8.8   How Many *Kanban* Do You Need?

procedures are repeatedly performed: in other words, standard operations. Before *kanban* can help maintain these things, however, we must establish an even spread of product models and volumes at the production planning stage.

At factories that include mostly standard, repeated operations, the number of *kanban* can be determined as shown below (see Figure 8.8), based on the premise of level production.

If the factory specializes in custom-order products, each order will need one *kanban* as the work order *kanban*. However, this *kanban* should also indicate when to produce the ordered item. And if, for example, the finished products at a certain process are placed into two or three different places, the *kanban* should also indicate from which site or sites the next process will withdraw the product. Even in this case of a custom-order factory, the *kanban* serves not only as a placement *kanban*, but also as an indicator of when the next process may come to withdraw items under a pull system.

Administration of *Kanban*

Kanban *Administration in Processing and Assembly Lines*

At one time, *kanban* was a big fad in Japan. It seemed that every factory was adopting the *kanban* system. But nine out of ten companies that adopted it found it did not work for them as they had expected. What was the problem?

Usually, the problem was that the factory tried to reap some benefits from the *kanban* system alone, without bothering to change its "shish-kabob" production system or its "push" system for moving goods through the line.

From the perspective of eliminating waste, it is best not to use any *kanban* at all. After all, for a factory to have *kanban*, it must have in-process inventory, and in-process inventory is itself a form of waste.

Unfortunately, the use of *kanban* can become a counterproductive fixed idea, just like any other firmly established practice. People eventually delude themselves into believing that their factory could not possibly operate without *kanban*. Before adopting *kanban*, it is best to take on the challenge of establishing thorough flow production.

Figure 8.9 shows an example of how *kanban* are used in assembly and processing operations. In this case, the transport *kanban* are the pallets themselves and the production *kanban* are hung on the "dispatch board" used for work scheduling.

After the improvement, this factory had sharply reduced its inventory levels compared to its previous days of production determined by the operations schedule. The factory was

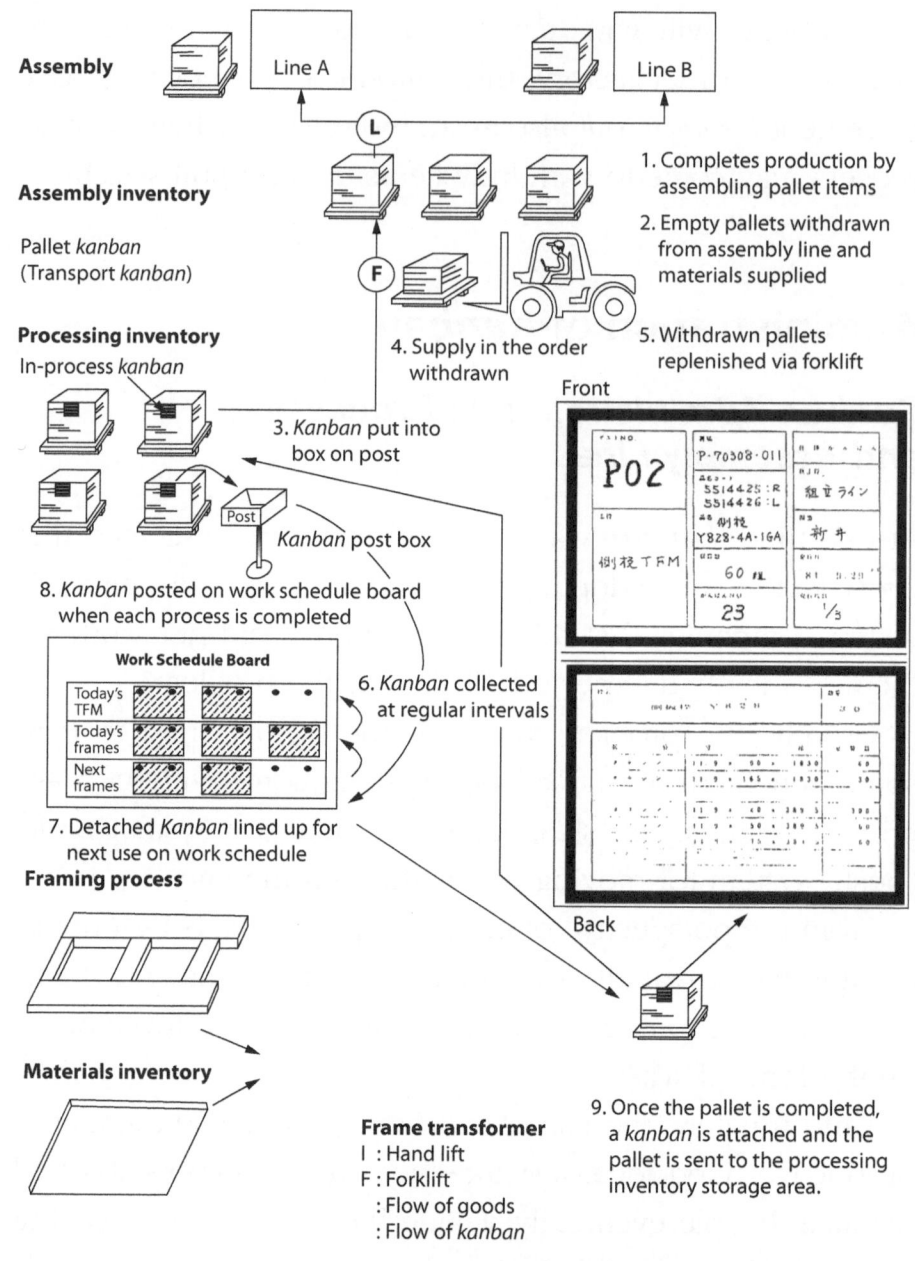

Figure 8.9   Use of *Kanban* in Processing and Assembly Lines.

also able to greatly reduce its lead-time for manufacturing scheduling and boosted productivity to about double its prior level. In addition, the flow of goods was made much more visible, which made problems easier to discover. Even when the required output rises, the factory is able to respond with faster turnover instead of larger lots, so it can maintain fairly steady inventory levels.

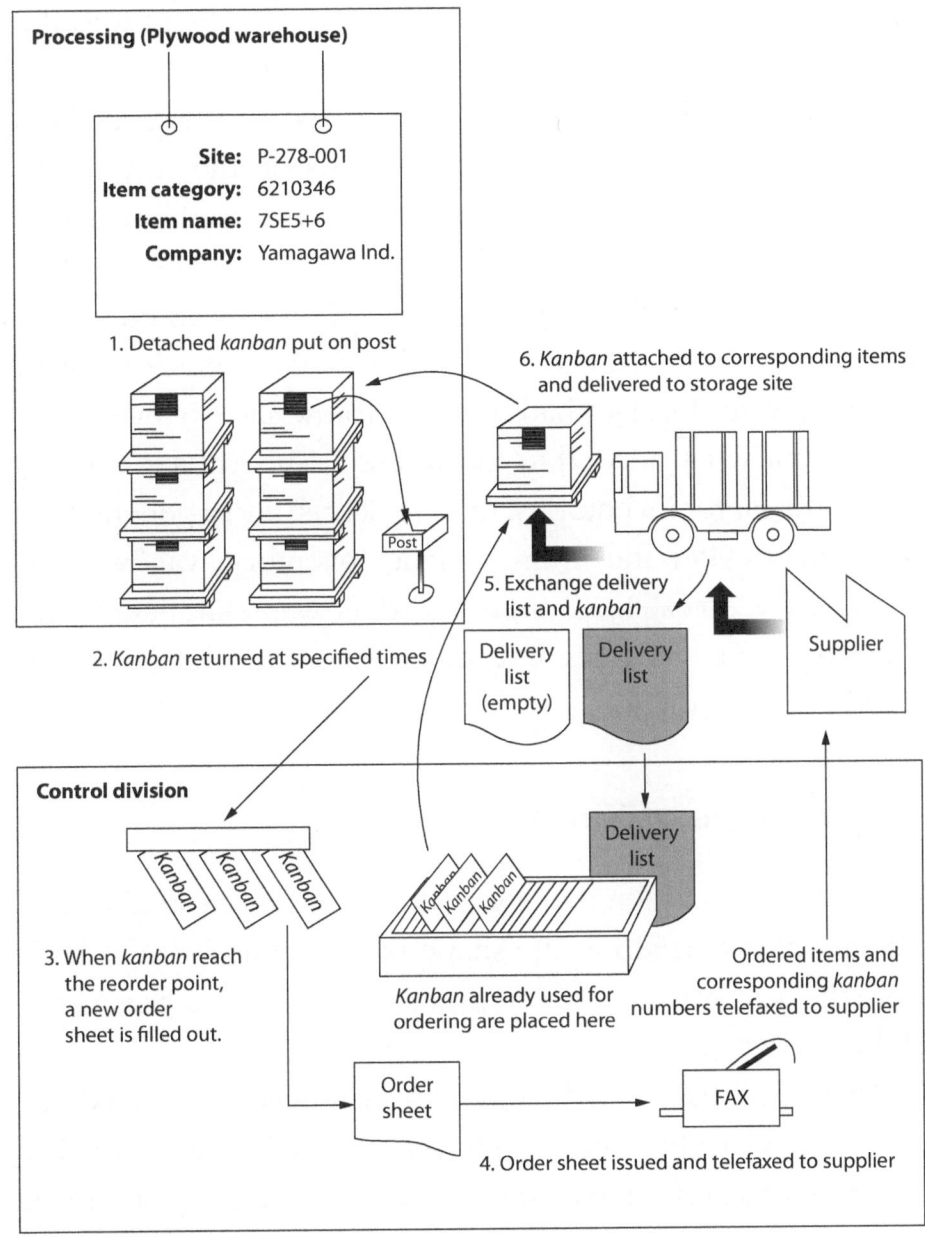

Figure 8.10 Example of Purchasing *Kanban*.

Administration of Purchasing-Related Kanban

Figure 8.10 shows an example of purchasing *kanban* that indicate information about withdrawn items. In this example, the *kanban* are not passed to the purchasing agent, but instead are used only in the factory. To make this possible, the factory counts the number of *kanban* to obtain the number of orders, then fills out an order sheet and telefaxes it to the purchaser, along with the individual *kanban* numbers.

When items are delivered, the *kanban* having those numbers are picked up and attached to the items on the way to the storage site. This means the *kanban* are also used in place of delivery vouchers.

Before making this improvement, the person in charge of ordering had no clear idea of how goods were flowing in the factory, and in fact had to come to the factory every day to find out what needed to be ordered. This situation led to larger and larger inventories, missing items, and a general lack of stability. After the improvement, inventory was reduced sharply, the problem of omitted orders was eliminated, and materials processing became much smoother thanks to the stable supply situation.

A Novel Type of Kanban

Figure 8.11 shows a rather exceptional and interesting example in which *kanban* in the shape of golf balls are sent back from the assembly line to the processing line via a pneumatic chute and gutter.

This "golf ball" *kanban* system eliminates the need for manually retrieving and issuing *kanban*. When an assembly line worker starts using a new box of parts, he or she removes

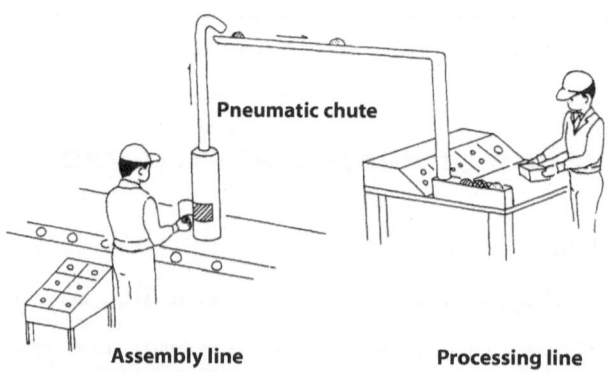

Figure 8.11   Example of Golf Ball *Kanban.*

the golf ball *kanban* that comes with the box and sends it through the chute back to the processing line.

These golf ball *kanban* indicate output amounts and use different colors to indicate different product models. The pneumatic chute places the balls onto a gutter that carries them across a distance of 200 to 300 meters. They are then "plunked" right in front of the processing workers. Since the golf balls come in the order in which the parts boxes are used on the assembly line, it is easy to maintain that same order on the processing line.

Chapter 9

Visual Control

What Is Visual Control?

Why Aren't Improvements Happening?

Many factories that are rich in improvement activities are poor in actual improvements. It is not so much that they do not know how to go about implementing improvement activities; it's just that they have failed to identify the factory's current problems and the various forms of waste that inhabit the place.

There are some excellent factories around, and there are some wretched ones. But the former do not necessarily have fewer problems than the latter. Every factory has lots of problems—not one is problem-free. So what separates the good factories from the bad ones? The answer is seen in the way they respond to problems: Good factories respond promptly and effectively, bad ones respond slowly (if ever) and ineptly. Good factories are good at revealing hidden problems. They are also good at getting the whole company behind finding the root causes of problems and making corrective improvements.

But things are never the same from one day to the next. No sooner have we solved yesterday's problems than we find today's problems staring us in the face. The question is, do we continue to jump at the opportunity of analyzing and solving the steady flow of problems as they arrive? If we do, our improvement activities are going somewhere.

Keeping up with problems as they occur is where bad factories fail. First of all, current problems and waste are often not easy to identify. The reasons why certain defects occur or why certain deliveries tend to run late are hidden or extremely vague. Factories tend to overlook such problems or "let them slide." Obviously, such an attitude will not bring success in solving problems and eliminating waste.

Even when a factory is successful in solving a set of problems, the world keeps changing. Before they know it, the factory employees have a new, perhaps more difficult, set of problems on their hands. The longer they are kept busy with those problems, the more time new problems have to accumulate. Eventually, the factory finds itself overwhelmed by the crush of problems and is no longer able to navigate the treacherous road to survival.

How can factories keep pace with the daily onslaught of problems? The answer is threefold:

1. By learning to distinguish promptly between what is normal and what is not.
2. By making abnormalities and waste obvious enough for anyone to recognize.
3. By constantly uncovering needs for improvement.

"Visual control" begins with making the factory's myriad abnormalities and forms of waste so clear that even a rookie will recognize them.

All too often, factory management becomes a desktop activity centered on statistics and number-crunching. Only the specialists understand what is going on with all those numbers. For example, let us consider what many factories do with their inspection results and other quality-related information. They take the numbers and plot them on various types of charts. And that's it. Rarely do they use such information as ammunition in improvement campaigns.

Types of Visual Control

Visual control is what JIT production offers as a means of turning specialist-knowledge management into plain and transparent management by everyone. We might even go as far as to say that visual control is JIT's way of "standardizing" management.

Visual control includes many application methods, each suited to a different type of management problem. Some visual control methods help identify waste while others help bring latent problems to the surface.

Figure 9.1 lists visual control's main tools and methods, which are described below.

1. Red tag strategy

   The "red tag strategy" refers to the red tags that are used when establishing the "5S's": proper arrangement (*seiri*), orderliness (*seiton*), cleanliness (*seiso*), cleanup (*seiketsu*), and discipline (*shitsuke*). The red tag strategy helps lay the foundation for improvement by making obvious which items are not needed for daily production activities.

2. Signboard strategy

   The signboard strategy is another visual control tool for establishing the 5S's. Signboards clearly show where tools and other items belong in the workshop so that anyone can find his or her way around easily.

3. White demarcators

   White tape or paint can be used effectively to enforce orderliness by marking off pathways, inventory storage sites, and other areas.

4. Red demarcators

   We use red demarcators on warehouse shelves, in-process inventory storage areas, and other inventory storage sites to indicate the maximum allowable amounts of inventory. In addition to using red marks to indicate maximum levels, we might also use green tape or paint

No.	Name	Illustration	Description
1	**Red tag strategy**	Red tag	The red tag strategy helps us distinguish needed items from unneeded items in the workshops. Red tag teams use red tags to mark unneeded items for removal.
2	**Signboard stragety**		In the signboard strategy, we set up signs that indicate what belongs where and in what amount, so that anyone will be able to understand where things belong.
3	**White line demarcators**		When organizing workshops in an orderly condition, marking out pathways and in-process storage sites with white tape makes it easy for anyone to keep the workshop neat.
4	**Red line demarcators**		Red line demarcators form part of the signboard strategy. We set up poles next to inventory (warehouse or in-process inventory) stacks and mark the maximum allowable stack height with a red line to show when excess inventory exists.
5	*Andon* **(alarm lamps)**		*Andon* immediately alert factory supervisiors to abnormalities that occur in the factory.
6	*Kanban*	Assembly	*Kanban* are administrative tools that help us maintain Just-In-Time production. The two main types of *kanban* are transport *kanban* and production *kanban*.
7	**Production management boards**		These are display boards that indicate current conditions on production lines. Data shown on these boards include production results, operating conditions, and causes for line stops.
8	**Standard operation charts**	Standard operation combination chart	We use these charts to find the work methods that use the best combination of people, machines, and materials. One of these charts should be on display at each line in the factory.
9	**Defective item displays**		Set-up at workshops where defects have occurred, these displays exhibit defective items along with graphic data urging workers not to allow the same defects to recur.
10	**Error prevention**	Error prevention board	Error prevention boards help promote independent management to reduce human errors.

Figure 9.1 Visual Control Tools and Methods.

to show minimum levels. The idea is to make inventory shortages or surpluses obvious for everyone.

5. *Andon*

As the "front line" leaders in the factory, supervisors such as foremen and section chiefs need to keep a close and steady watch on workshops to make sure the workers and the machines are doing the job right. When an abnormality occurs at a certain process, *andon* (alarm lamps) will alert the supervisors to the problem immediately.

6. *Kanban*

Kanban are an administrative tool that helps maintain the "pull" system and Just-In-Time production. The two main types are transport *kanban,* which are withdrawn whenever in-process inventory is withdrawn, and production *kanban,* which provide operation instructions at various processes.

7. Production management board

These boards show the current production line conditions. Besides showing estimated and actual output results, they indicate causes for line stops and various operation-related data. This keeps the line leaders constantly informed of the line's pace relative to estimated output. In other words, they always know if their line is going too fast or too slow.

8. Standard operation chart

Standard operation charts help us create easy-to-read graphical representations of process layouts, work procedures, and the like. In a sense, they serve as guide maps for those who prefer illustrations over descriptions.

Standard operation charts are rarely used by themselves. Usually, they are used with "standard operation combination charts," which help us find the most efficient combination of people, machines, and materials.

9. Defective item display

Quality control statisticians use Pareto diagrams to illustrate data on defective items and defect causes. Most factory workers, however, find it difficult to read Pareto charts. Defective item displays solve this problem by exhibiting actual defective items along with the Pareto diagram or other charts describing defect trends. (See Figure 9.2.)

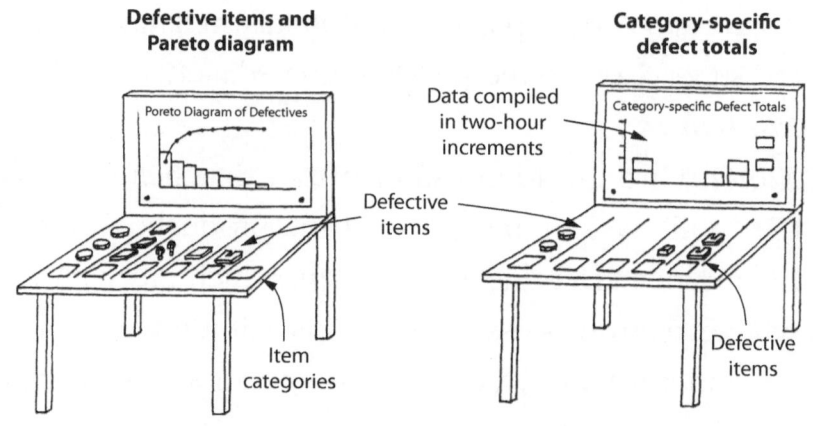

Figure 9.2   Defective Item Display.

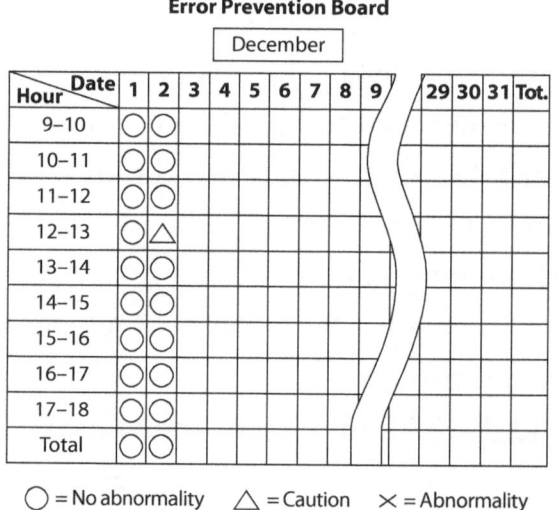

Figure 9.3   Error Prevention Board.

10. Error prevention board

Instead of remembering that "to err is human" and shrugging it off as inevitable, we can utilize error prevention boards to keep us more aware of our past errors so that we are less likely to repeat them. As such, error prevention boards are a tool for independent management. (See Figure 9.3.)

These boards usually have the hours of the day on the vertical axis and the days of the month on the horizontal axis. When workers receive feedback on defects or human errors from the next process, they mark the error prevention board under the hour and day when the defect occurred. They use one of three symbols to describe the type of defect or error: A circle indicates an error that does not cause an abnormality, an "X" indicates an error that does cause an abnormality, and a triangle serves as a caution symbol. At regular meetings, workshop leaders and workers review their errors and compare them to error prevention board results from previous months.

Case Study: Visual Orderliness (*Seiton*)

In Chapter 4 of this manual, we provided a detailed description of 5S-related visual control tools, such as the red tag strategy and the signboard strategy. Now we will examine a case study of how "Visual orderliness" (*seiton*) tools have been put to work.

First, let us reaffirm that orderliness means "standardizing where things go." In this case, standardizing means "making it clear to everyone what is normal and what is abnormal." With this in mind, let us see how well the 5S's were established in a parts storage area of a household electronics factory. (See the photo in Figure 9.4.)

On a scale of one to 100, these shelves rank about 25 for orderliness. Points were taken off for several reasons:

Reason 1: The shelves include place indicators, but no address indicators. What do the boxes' vertical arrangement signify? Their horizontal arrangement?
   Penalty: 15 points.
Reason 2: The boxes have item indicators but the shelves do not. How do people know where boxes should go on the shelves?
   Penalty: 15 points.

Figure 9.4   Establishing Orderliness in an Electronics Parts Storage Area.

Reason 3: The boxes give no indication of volume contained.
Penalty: 15 points.

Reason 4: There is not enough space above the boxes for us to easily see what is inside them. Perhaps the boxes are bigger than they need to be.
Penalty: 10 points.

Reason 5: The most serious reason is that the boxes can only be identified by the person who stocked them. This invites misplaced and lost items. It marks the beginning of the end of 5S conditions.
Penalty: 20 points.

Thus, by looking critically at the parts shelves and evaluating them based on the 5S's, we can more easily see where improvement needs exist.

Figure 9.5 shows a group of parts shelves at an automobile assembly plant. Let us compare these shelves with those shown in Figure 9.4 and note their differences.

Figure 9.5   Parts Shelves at an Automobile Assembly Plant.

Difference 1: The shelves at the auto plant are lower and thus accessible to shorter workers. Since the household electronics assembly plant hires more female workers than the auto plant, one would think it should have the lower parts shelves.

Difference 2: The parts boxes at the auto plant are smaller. This indicates that the turnover of parts boxes on the shelves is probably more frequent at the auto plant than at the household electronics plant.

Difference 3: The auto plant's shelves clearly show where each box goes, making them much easier to use than the other plant's shelves.

Difference 4: The location indicator signs at the auto plant are within the space marked off by white line demarcators, but they stick out beyond this boundary at the household electronics plant, which can be dangerous when tall items are being moved alongside the shelves.

Difference 5: Unlike at the household electronics plant, the parts boxes at the auto plant are easy to look into.

Difference 6: The biggest difference lies in how items are placed onto and retrieved from the shelves. At the auto plant, workers go to one side of the shelves to stock boxes and the other side to retrieve them. This results in parts being used in FIFO (First In, First Out) order.

The arrangement of shelves at the household electronics plant does not allow for FIFO stocking.

At first glance, one would not notice such differences between the two sets of shelves in the photos. After looking at them from the perspective of the 5S's, however, it is obvious that the auto plant's shelves are much more orderly than the household electronics plant's. With practice, we should all be able to make equally revealing evaluations at our own factories.

Standing Signboards

Kaizen *Boards*

Improvements tell the history of the factory and must keep pace with fast-changing market needs. Once we make an improvement, however, we begin to forget how conditions were before. It would be helpful indeed to keep track of improvements, so we can see how some improvements lead to other ones.

Figure 9.6 is a *kaizen* board that contains an "improvement results chart." Charts such as these can provide before and after displays for each improvement. By the way, it helps to take before and after photographs of the workshop from exactly the same camera position. Another way to enhance visibility is to choose a different "improvement color" each year and paint each improved workshop area using the improvement color designated for the year.

It is also good to include information such as improvement expenses and improvement descriptions in the displays.

Process Display Standing Signboards

Signboards are needed not only to show where things go, but also to describe machines and other equipment and show which processes are contained in processing and assembly lines.

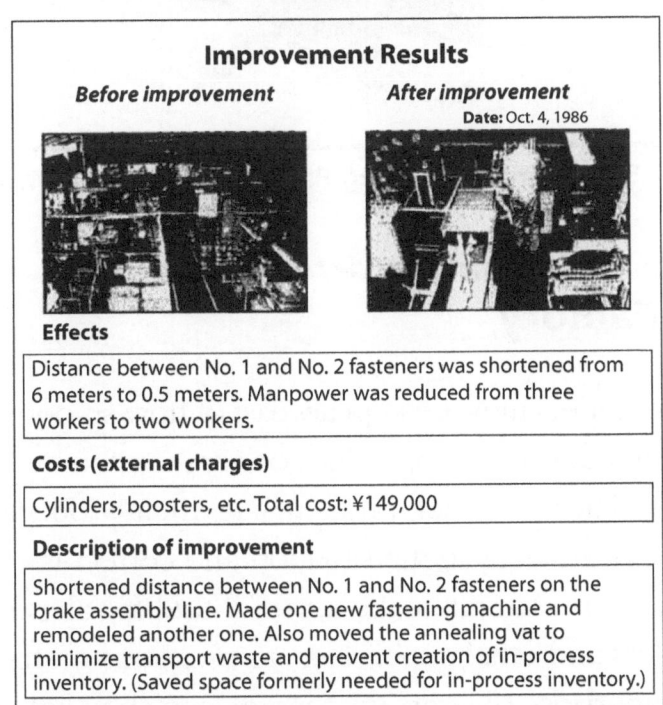

Figure 9.6 *Kaizen* **Board with Improvement Results Displays.**

Figure 9.7 depicts a signboard that describes the processes in a VCR assembly line. The signboards are posted alongside *andon* that alert supervisors to parts supply problems. In this case, the signboards serve a basic function in helping to ensure a smooth supply of parts to the assembly line with minimal errors or waste.

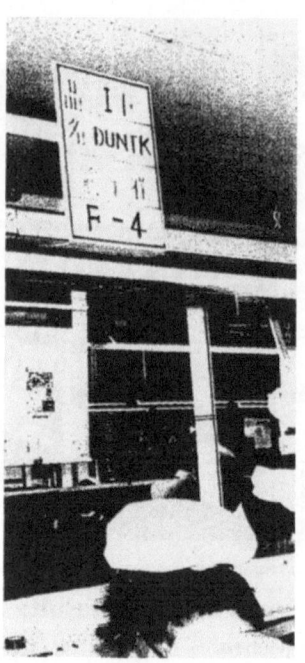

Figure 9.7   Signboards Indicating Processes in Assembly Line.

Andon: Illuminating Problems in the Factory

Workshop leaders must be kept abreast of how smoothly things are going in their workshops. The sooner they can be informed of abnormalities or other problems in their workshops, the sooner they can analyze the situation and correct it.

Andon (alarm lamps) make a useful tool for alerting workshop leaders and other supervisors to problems on the factory floor. The purpose of lamps in general is to shed light on dark areas. *Andon* are special lamps that illuminate problems in the factory.

Basically, there are four types of *andon*: "paging *andon*" that light up when supplies of parts are needed, "emergency *andon*" that notify supervisors of abnormalities, "operation *andon*" that indicate the equipment's operation status, and "progress *andon*" that confirm the progress of operations. (See Figure 9.8.)

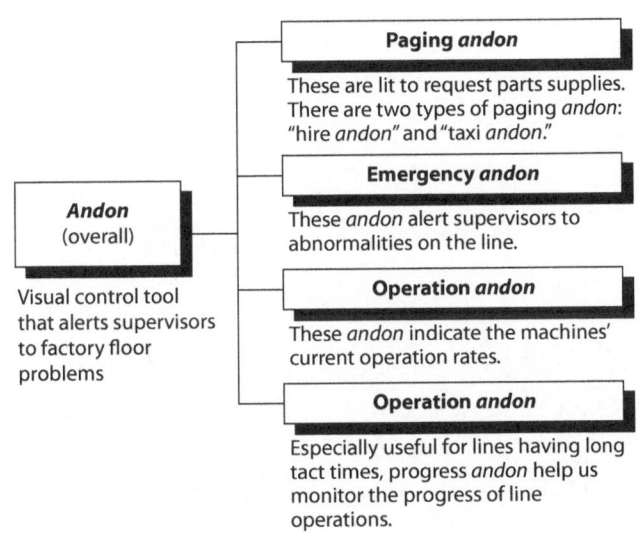

Figure 9.8 Types of *Andon*.

Paging Andon

Paging *andon* are often used to request supplies of parts for the production line. When parts are about to run out at a process, the operator sends out a signal that lights the *andon*. This notifies the people who operate the parts supply system. In Japan, the parts suppliers' quick movement around the factory in collecting and supplying parts has earned them the nickname *"mizusumashi"* or whirligig beetle.

Actually, there are two types of "whirligig beetle" techniques. One is the "hire" method, in which a group of *andon* page the carts used for supplying parts. The other technique is the "taxi" method in which dispersed *andon* page the carts.

Figure 9.9 illustrates the "hire" method for paging *andon*. In this case, the *andon* operate as follows:

Step 1: Operator confirms the shortage of parts and presses parts request button.

Step 2: The paging *andon* lights up.

Step 3: The parts supply cart operator (whirligig beetle) goes to the process where the parts request was issued.

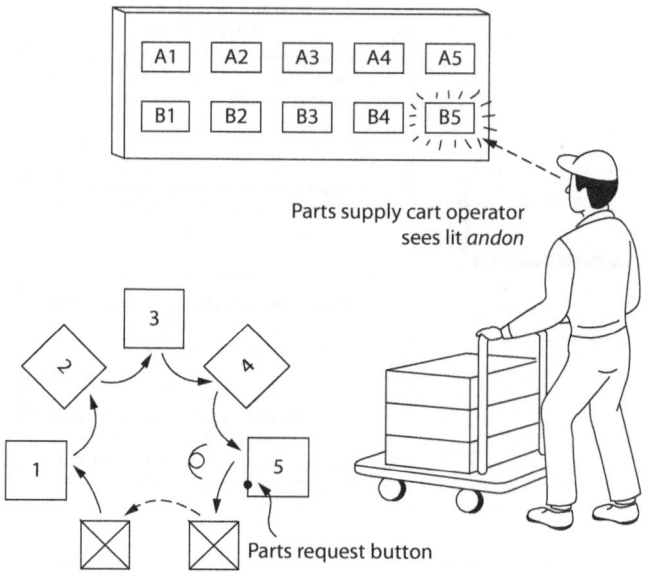

Parts supply cart operator sees lit *andon*

Parts request button

Figure 9.9   The "Hire" Method for Using Paging *Andon*.

Step 4: The parts supply cart operator takes the processes' empty pallets to the empty pallet storage area.

Step 5: The parts supply cart operator supplies the requested parts.

Step 6: The parts supply cart operator switches off the parts request button.

Warning Andon

Warning *andon* are mainly used on assembly lines and may differ depending upon the length of the line.

On short assembly lines, people tend to use "airplane *andon*." Like the flight attendant call buttons on passenger seats in commercial airplanes, each process in the assembly *line* has an *emergency* call button. When one of these buttons is pressed, the *andon* board for the assembly line lights up and shows which process's button was pressed. (See Figure 9.10.)

The following is a step-by-step description of how "airplane *andon*" are used.

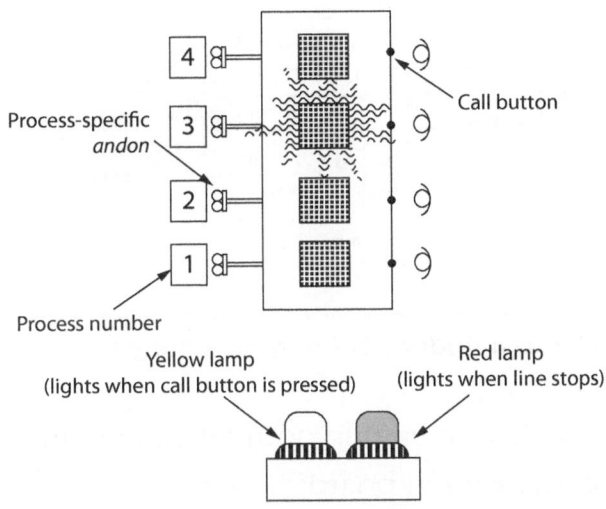

Call button

Process-specific *andon*

Process number

Yellow lamp
(lights when call button is pressed)

Red lamp
(lights when line stops)

Figure 9.10   Warning *Andon* for Short Assembly Lines.

Step 1: When a line operator gets behind due to parts shortages, defects, machine trouble, or whatever, he or she presses the "call button" (which lights up a yellow *andon* lamp).

Step 2: If the line is equipped with human automation devices for automatic stopping, the operators continue working for the time being. If it is not so equipped, a line operator must press the line stop button to stop the line, at which point a red *andon* lamp goes on.

Step 3: A workshop leader and/or a parts supply cart operator comes immediately to find out what the problem is, resolve it, and turn off the *andon* lamps.

The above type of warning *andon* configuration works well enough when the assembly line is short enough so that all of the *andon,* processes, and operators can be seen from one place. Longer lines, however, make it impossible to see the whole line and all of its operators. In this case, the *andon* are lined up in a centralized board (as in the "hire" method described earlier), as shown in Figure 9.11. These *andon* are used in three ways:

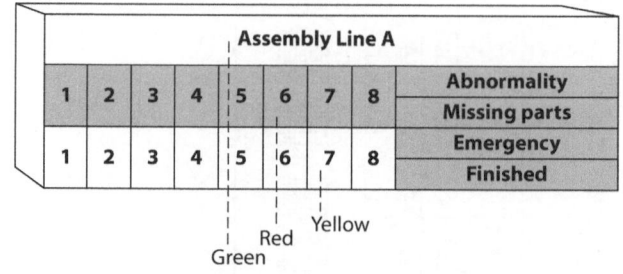

Figure 9.11   Warning *Andon* for Long Assembly Lines.

1. When line A is operating normally, the name "Line A" is lit up on the *andon* board.

2. When an abnormality occurs, an operator presses a call button, at which time the process number where the call was issued lights up on the *andon* board (usually a yellow lamp).

3. Once a warning call button is pressed, if the line is equipped with a device that automatically stops the line at a certain point, the line will continue until that point is reached or until the problem is resolved (whichever comes first). If the line is stopped, the yellow *andon* indicating the process number goes out and is replaced by a red *andon* that also indicates the process number.

Operation Andon

Operation *andon* indicate machine operating statuses. When the machine has been stopped, the operation *andon* shows the reason for the stoppage. (See Figure 9.12.)

Operation *andon* can be used as follows:

1. The green "IN OPERATION" lamp is lit whenever the machines are operating normally.

2. The yellow "CALL" lamp is lit when an emergency call button has been pressed.

3. A red lamp ("BREAKDOWN," "BLADE CHANGE," or "WIDTH ADJUST") is lit when a corresponding button has been pressed.

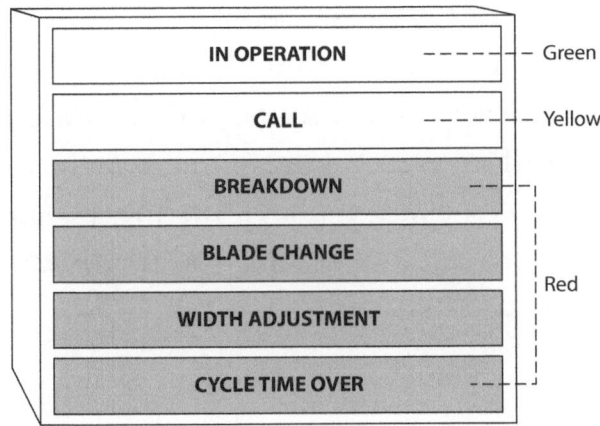

Figure 9.12   Operation *Andon*.

4. The red "CYCLE TIME OVER" lamp is lit when one of the cycle time pacemakers installed in the equipment indicates the cycle time has been exceeded.

Progress Andon

Many assembly lines have short pitch times, such as 1- or 2-minute tact intervals. When a line has such a short tact time, the progress of operations is easy to observe simply by monitoring the rhythm.

It is more difficult to sense delays in lines that have longer tact times, such as 10 or 20 minutes. Progress *andon* enable line operators to gauge the progress of their own operations. (See Figure 9.13.)

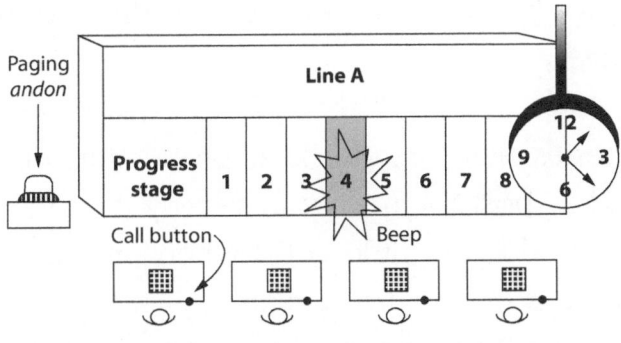

Figure 9.13   Progress *Andon* in Manual-Conveyance Assembly Lines.

Most progress *andon* are divided into 10 equal sections, but the main idea is to have the various stages of the operations correspond in a level manner to the various *andon* sections.

In the case of conveyor lines, limit switches can detect when the progress stage number on the progress *andon* board has changed. In lines where workpieces are passed by hand, a timer is used instead.

Production Management Boards: At-a-Glance Supervision

Many factories rush feverishly into production based on the production schedule and resort to overtime hours if it turns out they cannot keep up with the scheduled output. To help avoid such unpleasant surprises, it would be nice to know from one hour to the next just how the line is doing, whether it is too slow (and why), and what countermeasures to take under various circumstances.

Production management boards serve exactly this purpose.

Production management boards should be simple in design and should emphasize providing information that answers the following key questions:

■ How do current results compare to estimated results?
■ Why was the line stopped the last time?
■ What kind of improvement is needed?
■ Will there be any spillover into overtime or tomorrow's schedule?

The factories that already have production management boards tend to post them only in factory managers' offices. However, they do very little good when only the managers can keep an eye on them.

The people on the factory floor—the workshop leaders and equipment operators—have the greatest need to have

Production Management Board for

| December 1 |

Line A	Cycle time: 60"			Cycle time			Comments
Hour	Est.	Act.	Diff.	Est.	Act.	Diff.	
8:30–9:30	60 / 60	58 / 58	–2 / –2				*Defect at Process No. 1*
9:30–10:30	60 / 120	60 / 118	0 / –2				
10:40–12:00	80 / 200						
12:45–2:00	75 / 275						
2:00–3:00	60 / 335						
3:30–4:00	60 / 395						
4:00–5:00	60 / 455						

Figure 9.14   Production Management Board.

production management boards to keep them informed. It is a good idea to have a production management board posted as "the final process" in the line, so that everyone checks it at least once per production cycle. Nothing works better to keep workshop leaders and operators aware of current conditions in their workshops and conscious of problems and their solutions.

Most production management boards look something like the example shown in Figure 9.14.

Relationship between Visual Control and *Kaizen*

This topic reminds me of a visit I once paid to a European automobile assembly plant. While touring the plant, I noticed a large and fancy *andon* hanging from the ceiling at the final process in the assembly line. They must have spent a lot of money to buy and install that *andon*, much more than any Japanese manufacturer would have spent. I also noticed, however, that it never seemed to light up at all.

Curious, I asked the factory's chief production engineer why the *andon* was not lit.

He said that when the company managers went to visit some factories in Japan, they were very impressed with the *andon* there and decided to adopt the tool in their own factory. Once the *andon* was installed, the line workers soon found that no one ever came to the rescue when they pressed the "call" buttons, and so the problems that prompted them to call for help did not get resolved. In fact, the only change the *andon* made was to create the wasted motion involved in pressing the call buttons!

Within a month of the *andon's* arrival, the workers stopped pressing the call buttons, and eventually it was decided just to unplug the *andon* to save electricity costs.

Nonsensical as it sounds, this case was not an isolated oddity. Similar episodes have occurred in America and even in Japan.

All too often, people have casually adopted the external trappings of JIT production, such as the various JIT tools and techniques, without committing themselves to learning the concepts and spirit of JIT. The results of such misguided approaches include wastebaskets full of *kanban,* completely baffling standard operations that lack any trace of rationale, and decorative *andon* that hang from the ceiling like ill-conceived chandeliers.

No matter how many visual control tools we bring into the factory, if we do not use them correctly to discover and promptly correct abnormalities, the tools are no more valuable than money that is always kept under a mattress.

If we can make abnormalities obvious and perform prompt analyses of their causes, we can expect to make improvements based on such discoveries and analyses at least half of the time.

Figure 9.15 shows the roles that various visual control tools can play in the improvement cycle. Let us remember that *just introducing visual control tools will not automatically result*

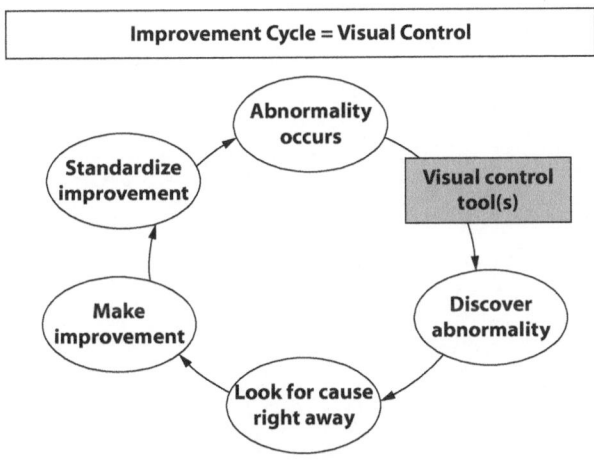

Figure 9.15   The Roles of Visual Control Tools in the Improvement Cycle.

in improvements. Visual control tools only do one-third of the job—they help make abnormalities obvious and therefore easier to discover. The other two-thirds—analyzing the abnormality and taking corrective action—still must be done by the factory people themselves.

Visual control is meaningless unless we look at it from the larger perspective of the improvement cycle.

Index

A

K

About the Author

Hiroyuki Hirano believes Just-In-Time (JIT) is a theory and technique to thoroughly eliminate waste. He also calls the manufacturing process the equivalent of making music. In Japan, South Korea, and Europe, Mr. Hirano has led the on-site rationalization improvement movement using JIT production methods. The companies Mr. Hirano has worked with include:

Polar Synthetic Chemical Kogyo Corporation
Matsushita Denko Corporation
Sunwave Kogyo Corporation
Olympic Corporation
Ube Kyosan Corporation
Fujitsu Corporation
Yasuda Kogyo Corporation
Sharp Corporation and associated industries
Nihon Denki Corporation and associated industries
Kimura Denki Manufacturing Corporation and associated industries
Fukuda ME Kogyo Corporation
Akazashina Manufacturing Corporation
Runeau Public Corporation (France)
Kumho (South Korea)
Samsung Electronics (South Korea)
Samsung Watch (South Korea)
Sani Electric (South Korea)

Mr. Hirano was born in Tokyo, Japan, in 1946. After graduating from Senshu University's School of Economics, Mr. Hirano worked with Japan's largest computer manufacturer in laying the conceptual groundwork for the country's first full-fledged production management system. Using his own interpretation of the JIT philosophy, which emphasizes "ideas and techniques for the complete elimination of waste," Mr. Hirano went on to help bring the JIT Production Revolution to dozens of companies, including Japanese companies as well as major firms abroad, such as a French automobile manufacturer and a Korean consumer electronics company.

The author's many publications in Japanese include: *Seeing Is Understanding: Just-In-Time Production (Me de mite wakaru jasuto in taimu seisanh hoshiki), Encyclopedia of Factory Rationalization (Kojo o gorika suru jiten), 5S Comics (Manga 5S), Graffiti Guide to the JIT Factory Revolution (Gurafiti JIT kojo kakumei)*, and a six-part video tape series entitled *JIT Production Revolution, Stages I and II*. All of these titles are available in Japanese from the publisher, Nikkan Kogyo Shimbun, Ltd. (Tokyo).

In 1989, Productivity Press made Mr. Hirano's *JIT Factory Revolution: A Pictorial Guide to Factory Design of the Future* available in English.